As our wastes accumulate &
changing in air ...
materials. It ...
unsuitable ...
habitation, and possibly even
for industrialization

The
Economics
Of
Environmental
Protection

DONALD N. THOMPSON
York University, Toronto, Ontario

Winthrop Publishers, Inc.

Cambridge, Massachusetts

Library of Congress Cataloging in Publication Data

Thompson, Donald N
  The economics of environmental protection.
  Includes bibliographical references.
    1. Pollution—Economic aspects—United States.
  I. Title.
TD180.T49   338.4′7′36360973   72–11520

Copyright © 1973 by Winthrop Publishers, Inc.
    17 Dunster Street, Cambridge, Massachusetts 02138

Contents

iii

Preface

The future of the earth is in our hands.
How shall we decide?
—Teilhard de Chardin

The two sciences of economics and ecology, whose names derive from the same Greek word, have had remarkably little contact with one another—perhaps because of the way in which we departmentalize and compartmentalize the teaching of each. Sociologists, political scientists, and certainly earth scientists have had far more contact with ecology than have economists. Where the two sciences have met, ecologists have tended to treat economics as a subset of ecology, and vice versa. Economists tend to view ecological questions as variables which can be inserted into their models, a view which does not induce economists to change their basic models but merely to accommodate them to new variables.

This book is not intended as a pedantic application of microeconomic theory and welfare economics to pollution problems. Rather, it is an economist's view of environmental protection rather than a formal discussion of any underlying economic theory. To the extent that economic theory and models are introduced, it is hoped that the material can have equal value to students of economics and of ecology, and be of some interest in a number of other disciplines.

The problems that distinguish pollution from other resource allocation areas include the existence of economic externalities; for instance one man's noise affects another man's sleep, and the non-localized effects of problems like air pollution. Such environmental situations are simultaneously simple and complex problems. For instance, the dumping of mercury into lakes and streams is a relatively simple economic problem. Mercury is a deadly poison to fish in very small amounts and to humans in larger amounts. On the abatement side, economists are faced with relatively small costs of elimination. The outcome is usually a fairly swift legislative response (pass a law), judicial response (issue an injunction), or economic response (require polluters to compensate victims and their heirs for the full cost of pollution damage).

vii

At the other extreme is the complex problem of controlling the pollution of waterways serving densely populated areas. Once we have eliminated deadly pollutants we have to deal with a whole range of industrial and domestic wastes (sewage). A flowing river normally carries enough dissolved oxygen to break down and assimilate a given quantity of degradable pollutants. Beyond that amount a small degree of pollution may bother fish, birds, and swimmers; a slightly larger degree may destroy a percentage of anadromous fish like salmon. A quantity of pollution which does no damage when diluted during periods of spring runoff may be sufficient to kill all marine life in periods of low water flow during late fall. In such situations, should society (or an individual polluter) spend $5 million to protect $2 million worth of fish? Should all polluters be required to abate their wastes equally, or proportionately? Should only the worst polluters abate, or only the latecomers? Or only those who can abate least expensively? These sorts of problems, which are considered in the text, are complex rather than simple, and the correct legislative, judicial, or economic response to them is not always intuitively clear.

This book takes a variety of approaches to the pollution topics discussed. Although some economic techniques and tools are mentioned in relation to some problems and not others, such a narrow approach is not inevitable. One of the great needs of the whole environmental protection field is to develop novel approaches to old problems. For example, compared with the available ways of dealing with water pollution, the economically feasible means so far applied to the problem of air pollution have been extremely limited. This is true both because it is easier to control water flows and contamination than it is to control meteorological events, and because air pollution can really only be treated at its source, while there are various alternatives with water pollution.

A number of environmental problems are not covered in this book; they are omitted not because they lack importance, but because those that are covered hopefully illustrate the nature of the problems, and some approaches to economic solutions. For example, the automobile-highway problem is not considered here, although it is analogous in many ways to the waste-disposal problem in that it needs a systemic approach to its solution. The massive and continuing highway construction program triggers a cycle of building more roads because people buy more cars; then building and selling more cars because there are more and better

highways. The result is a gross consumption and waste of resources, air pollution, noise pollution, human dislocation, and the
destruction of both urban and rural areas. The single-minded
emphasis on highways also effectively minimizes the emphasis on
alternative forms of ground mass transportation.

The reader of this book, who is probably white and middle-class,
should be aware that whatever the importance of ecology for the
future of man, at present the subject is very close to being a white,
middle-class issue for white, middle-class Americans. Where, in
the scale of priorities, is air pollution in the mind of a mother in a
central city whose baby has been bitten by a rat? What priority
does a polluted lake have to a family whose only recreation area
is a trash-filled alley? What priority does decontaminating mercury-
laden swordfish have to a child who hasn't enough of anything to
eat? What priority do open spaces occupy in the minds of thousands of human beings who are born, live, and die in congested,
dangerous slums? When ecologists talk of improving the "quality
of life" by reducing pollution, we must always be aware of whose
life we are talking about. And we must also be wary of becoming
so involved in saving the countryside that we allow the cities, and
the people in them, to decay.

The unity of subject matter in the book is such that only eight
chapter divisions have been used, but these need not be read in
order—with the exception that Chapters One or Five should precede Chapter Six for an understanding of the concept of external
costs.

All chapter sections are numbered to facilitate reference, cross
reference, and general use of the book. This is not a common
practice in the social sciences, although it is widely used in areas
such as law.

I wish to acknowledge my debt to a number of people who have
influenced the shape and content of this book. In particular, I
profited greatly from the class materials and discussion in a course
in Environmental and Consumer Protection organized by Profes-
sors Louis L. Jaffe and Lawrence Tribe of the Harvard Law School
in the fall of 1970. Much of their thinking is reflected in the arrangement of this book. The structure and part of the content of
Chapter Six had their origin in a paper prepared by Barry R.
Furrow, my research assistant at the Harvard Law School during
1970-71. I am indebted for many insights to a scientist's view of
these problems to Lee Hofmann of the Woods Hole Oceanographic

Institute, who read the manuscript during the summer of 1971 with an eye to the sort of mistakes that an economist makes in a multi-disciplined work. My thanks go also to Ardeen R. Thompson, Marshall I. Goldman, Rob Wilson, and Bruce Goldstein for their many helpful comments on the manuscript, to James Murray and Muriel Harman of Winthrop Publishers, Inc. for their encouragement in the face of deadlines violated, and to Noreen Maxwell, an able secretary who labored through numerous versions of the manuscript.

It would be unfair not to mention a number of people on whose research and writing in the areas of economics and ecology I have particularly relied. With apologies to all those I've probably overlooked, they include Kenneth E. Boulding, Blair T. Bower, Barry Commoner, J. H. Dales, Paul R. Ehrlich, Marshall I. Goldman, Robert Heilbroner, Hazel Henderson, John R. Henning, Milton Katz, Allen V. Kneese, Lester B. Lave, Edwin S. Mills, Richard Morse, Ronald C. Peterson, Ronald G. Ridker, Joseph Sax, Eugene P. Seskin, Azriel Teller, and Harold Wolozin.

Thanks are due to E. J. Mannion, Publisher of *The Canadian Magazine*, for permission to reprint the environmental poetry in Chapter Two from the June 19, 1971 issue.

Financial help with preparation of the manuscript came from a Ford Foundation grant to the Faculty of Administrative Studies at York University in Toronto, and from the Harvard Law School. Both assistances are gratefully acknowledged.

Donald N. Thompson
Cambridge, Massachusetts
and
Toronto, Ontario
October, 1972

1

Economics

And Environmental

Protection

Aboard one of the U.S. Navy's deep submersible
craft, fifty miles off the coast of San Diego,
and 2,450 feet down, Admiral R. J. Galanson,
chief of naval materials, peered through the
portholes to view the wonders of the undersea
world which perhaps no other man had ever
seen. The first thing he spotted, only two
feet away on the ocean floor, was an empty
beer can.

State Senator Leslie Robinson,
on Earth Day, May 22, 1970

100. The Pollution Crisis

The startling rate at which environmental pollution
has decended upon the world is illustrated by Alvin Toffler's "ac-
celerative curve" which is discussed in his book *Future Shock*.[1] If
the past 50,000 years of man's existence were divided into lifetimes
of 62 years each, there have been about 800 such lifetimes. Of
these 800, about 650 were spent in caves. Only during the past
70 lifetimes has there been written communication from one lifetime
to another. Only in the past 6 lifetimes have more than a few men
of any generation seen a printed word. Only in the past 2 lifetimes
has anyone anywhere used an electric motor. The vast majority
of all the material goods man uses today have been developed
within the present lifetime, the 800th. And, virtually all the pollution man is concerned with today—in the air, water, and in ecolo-

gical imbalances—is the product of the last lifetime; in many cases, of the last one-quarter lifetime.

The most extensive and concentrated pollution has taken place in the cities. Man is now undergoing the most rapid urbanization in his history. In 1850 only four cities on earth had a population of one million people. By 1960 there were 141 cities with a million people each. By the year 2000, there will be over 500 cities with populations of over a million people. The earth's urban population is increasing at a rate of 6.5 percent per year; it doubles every eleven years. Every eleven years we have to build another New York, another London, another Calcutta, another Shanghai, another Moscow, just to stay even. Were the rate of growth to level off, in thirty years we would need five to six times the urban facilities that existed in 1970.

But population is only one factor of the pollution problem; technology is the other. Despite the fact that we are already operating from a substantial industrial base, the rate of increase of the world's industrial production is staggering. In France in the 29 years between 1910 and World War Two, industrial production rose about 5 percent. In the 17 years from 1948 to 1965, it rose 220 percent. In the 21 western, industrial countries of the Organization for Economic Cooperation and Development (OECD), average annual growth rates for gross national product in the 1960's were just under 5 percent. Japan averaged 9.4 percent. All of Africa averaged about 6 percent. This implies a doubling of the total output of goods and services in the world every 15 to 18 years—and the doubling times are shrinking. Twice as many iron foundries, twice as many pulping mills, twice as many oil tankers, every fifteen or so years; and four times as many by the end of this century.

Much of the recent literature on economics and the environment has discussed the relative contributions of evolving technology and of increasing population to environmental problems. Both factors turn out to be of great importance; their interrelated status is discussed in the following section.

101. Technology or Population?

Because environmental problems have increased many hundred percent since 1940 while population increased only forty-five percent, population growth is often erroneously assumed to be a minor factor in the ecological crisis. Actually, the two factors are strongly interrelated; Paul Ehrlich claims that only fifty million Americans, given our existing technology, could eventually destroy the planet. It is difficult to imagine any American life style which would prevent three or four hundred million people from accomplishing the same end in the long run.

With increasing affluence comes a dramatic rise in production levels and resource usage. In 1940, each American used about 240 pounds of paper; by 1965, usage had more than doubled to 515 pounds. By 1970, it had increased to 700 pounds—almost three times the 1940 per capita figure. In 1940, U.S. electric utilities consumed 54 million tons of coal and fuel oil and released one million tons of sulphur oxides into the air; by 1968, utilities used 277 million tons of coal and oil, and released 11.5 million tons of sulphur oxides. Consumption of coal and oil to produce electricity rose from 4/10ths of a ton per person in 1940 to 1 4/10ths of a ton per person in 1968.

Similarly, the amount of nitrogen used in chemical fertilizers on farms (and washed off the fields to pollute ground and surface waters) increased from .42 million tons in 1940 to 5.9 million tons in 1968, an increase of fourteen times. The production of plastics, many of which are almost indestructible, rose almost four times between 1960 and 1970, from 32 pounds to 115 pounds per capita.

If we make the assumption that the growth of production (or of production of certain products like plastics, or production using electric power) is the underlying cause of pollution, then the only way we can get less pollution is by less production, either by having more of the labor force idle through more of the year, or by cutting economic growth by deliberate unemployment or deliberate production inefficiency. Since both solutions, plus the objective of restricting production, are not acceptable to most people in the short run, the alternatives left are to change radically the nature of things produced—e.g., fewer indestructible plastics—or to change power sources, or to concentrate on recycling.

Rising production, and the ideal of a critical mass in productive capacity, only begins to explain why we are being overwhelmed by pollutants now, but were hardly bothered two decades ago. The other factor to be considered—probably the key factor in talking about technology—is the effect of compound growth.

In 1945, when pollution was hardly talked about (except in

Pittsburgh), the gross national product stood at $350 billion; by 1957, it had grown to $450 billion. By 1969, after another 12 years, it had grown to $750 billion in terms of 1945 dollars. The GNP increased $100 billion in the first period, but $300 billion in the second period of equal length. This did not result from a more rapid growth rate, but from the law of compound interest. A GNP growing at 4 to 5 percent per year doubles in less than 20 years. Twice 100 is an increase of only 100; twice 200 is an increase of four times the original amount, and so on. The real output of goods and services in the United States has grown as much since 1950 as it grew in the 330 years from the landing of the Pilgrims in 1620 to 1950. In the next 12 years the GNP will rise about $550 million in constant dollars, given only a conservative growth rate of 4 to 5 percent per year. This represents an increase in 12 years of more output than the country achieved in the first 350 years of its existence. If the consequence of production is increased pollution, then a similar compounded rise in environmental problems can be expected to occur.

The population side of the problem is much more complex than some of the discussions (of zero population growth, for example) have indicated. If we assume that reduction in fertility within any national group ought to be voluntary and not imposed by the government, then population control must proceed through steps which motivate families to want fewer children, and must teach the use of available contraceptive methods. A voluntary reduction in population growth could be accomplished only after a delay of 20 to 30 years.[2] If worldwide efforts to limit population were begun immediately and were successful, world population would stabilize about the year 2000. Barry Commoner has estimated that the earth's population at that time will be from 6 to 8 billion, or about twice the present world population.[3] Some observers estimate that it is theoretically possible for overall world food production to keep pace with population growth of this magnitude to the year 2000; others have concluded that no foreseeable extension of food technology, with whatever side effects, can support a doubling of world food production.

The U.S. Bureau of the Census estimated in 1970 that the U.S. population would grow by 75 million people by the end of the century—an increase of 35 percent. Such a U.S. increase would be as significant as the overall world increase. Biologists like Paul Ehrlich pointed out that each American has roughly 50 times the negative impact on the ecology as the average citizen of India.[4] On the assumption that Americans do not pursue even more environmentally destructive activities in the future, an increase of 75 million Americans would be equivalent to adding 3.7 billion Indians to the world population. From the standpoint of consumption of scarce fuel and mineral resources, 75 million additional Americans are the equivalent of 2 billion Columbians, 10 billion Nigerians, or 22 billion Indonesians.

The future, however, may not be as dismal as the U.S. Census prediction would indicate. National birth statistics for the first quarter of 1972 showed that U.S. births had dropped sharply, even from an already declining 1971 rate. The total fertility rate—a sophisticated demographic measurement—had in late 1972 reached virtually the replacement level, which is 2.11, or the number of children women must bear, on the average, to replace themselves, their husbands, single people and women who die before bearing children. The late 1972 replacement rate was about 2.14, the lowest in American history.

Reaching the threshold to zero population growth is not the same as no growth, however. For the population to stabilize firmly, very low fertility rates would have to persist for some 30 years. No authority believes that this situation is even conceivable. For example, one need only examine the abrupt fluctuations in recent American replacement rates: 2.23 children in 1937; 3.27 in 1947; 3.77 in 1957; and 2.57 in 1967 to see how difficult stabilization would be. At the same time, the present low fertility trend cannot be dismissed as a fad; it is just as strong as the social forces that explain it, and these appear entrenched: the accelerating use of contraception, abortion, and sterilization; more single people and later marriage; more working wives; increasing concern for overcrowding and pressure on the environment. Similar low and declining birth rate trends are being experienced in virtually all western and northern European nations and in a wide assortment of developing nations.

Similar limits of growth may also be in sight for much of the production and technology in developed countries. The heaviest environmental pressures come from agriculture, mining, and heavy industry, each of which has developed to a mature stage in most western countries. Also the total volume of manufactured goods will probably increase at a declining rate simply because the time available to use all the things produced (a fourth car?) is becoming more scarce, and because an advanced economy grows by expanding in areas like education, communications, and finance, which burn little fuel and consume relatively small quantities of minerals. These trends are tentative and do not constitute grounds for great optimism; they do indicate, however, that our prospects may not be hopeless.

Growth may also be limited because the developing countries may not be able to industrialize along the lines of the more developed countries. For instance, the earth does not contain enough metal resources for every developing country to become an industrial power like the U.S. Even if there were enough resources, global industrialization would produce intolerable pollution. If all the presently underdeveloped countries were to achieve the economic level of the United States it could mean an increase of 200 times the present natural resource and pollution load on the world environment. Given the existing destruction of land, air, and especially the oceans, there is little reason to anticipate that any pollution control technology will be forthcoming in the near future which will be capable of handling such an increased standard of living even for the present population of the world.

Paul Ehrlich has suggested one answer: that some areas of the world be maintained indefinitely as charitable wards of the developed countries, with their people "maintaining their traditional ways of life," and provided by the developed countries with "access to the fruits of industrial societies," including medical services, educational facilities, teachers, and technical assistance in population control.[5] Such a policy might be hard to sell; persuading the people of the poor countries that they must forever remain as second-class citizens in world society ranks with convincing the people of China that they should give up their production of nuclear weapons and intercontinental missiles and be content with a role as an agrarian nonpower.

102. Who Pollutes?

The movement to reduce pollution and to protect natural resources is usually considered to be anti-industry. However, while the role of industry in the creation and dispersal of pollutants inevitably brings it into conflict with environmental protection movements and agencies, industry as such is only responsible for a portion of the pollution in the environment.

In the purest sense, all men are polluters, the rich not necessarily more than the poor. All automobile drivers foul the air with exhaust fumes; the 1963 Chevrolet much more than the 1973 Cadillac. Given our obsolete zoning laws, the homes of the poor intrude more rudely upon nature than the homes of the rich.

While the increasing inability of American rivers to cleanse and renew themselves has been caused in part by effluents from large industries and power generating stations, untreated sewage from towns and cities has been a more important source of damage. More important still in many cases has been the runoff from thousands of small farms, where honest farmers, seeking greater crop yields, have overused nitrogen fertilizers. The result has been eutrophication, a cycle whereby algae feed upon the fertilizers draining into the rivers and where decaying algae consume so much oxygen from the water that bacterial action which would have broken down organic wastes becomes nonexistent.

Some of the worst polluters are governmental bodies established for beneficial purposes. For instance, the Tennessee Valley Authority (TVA), set up to improve the economy of the Tennessee Valley, has indirectly aided in the devastation of the Kentucky mountains and the deterioration of the quality of life in that area. The TVA is the major customer of the coal mining companies that strip mine in the Kentucky mountains, leaving unreclaimed an area large enough to make a swath a mile wide extending from Times Square to the Golden Gate and back again almost to the Nevada border. What remains after strip mining in Kentucky is for the most part bare land gradually eroding and washing away to fill local streams with soil and acid mine drainage.

Some governmental bodies established to protect the public by regulating business activities have ended up by protecting offending corporations. In October 1969, the Federal Water Quality Administration requested stricter effluent control for a DDT plant at the Redstone Arsenal which was operated by Olin Chemical Corporation. The DDT-laden effluents from the plant were stated to be polluting the Wheeler National Wildlife Refuge and a reservoir from which the City of Decatur, Alabama drew its drinking water. The Department of the Army intervened, overruling the recommendation of the Federal Water Quality Administration on the basis of defense priorities, and allowing the alleged DDT effluent pollution to continue unchecked.[6]

RECURRING ISSUES IN THE ECONOMIC LITERATURE

103. Economic Externalities

The underlying economic dilemma of environmental protection is that it sometimes pays to pollute. It costs less for a manufacturer to dump harmful wastes into the air or the water than to take the effort to process them. It costs less to produce electricity if the power company can use the air and water to absorb waste gases and heat. It costs local officials less in the way of lost votes from higher taxes to not build a sewage treatment plant, than to continue to pollute the water used by towns and cities downstream. It costs a houseowner less to burn his grass clippings in a back yard incinerator than to pay to have them hauled away. To the polluter, pollution may mean lower costs or higher profits; to employees and customers, it may mean higher wages or less expensive products.

Why does the economic problem exist? What dysfunctions in our economic system encourage these things? It seems that the "cost" (or sometimes benefit) of polluting activities to society carries no price tag and is not considered by the private decision-makers who are responsible for their existence. The effects are known as economic externalities (and sometimes as spillover costs or side effects). Externalities such as air pollution and sonic booms from supersonic aircraft impose large social costs that are not private costs for the polluter, or for the airline. The resultant inequality of social costs and decision-makers' private costs indicates that economic resources are being improperly allocated in society. For instance, if high-sulphur fuel is cheaper to produce than low sulphur fuel, it will be burned and sulphur dioxide wastes will be dumped into the air. Society pays a price in terms of damage to paint, metal surfaces, plants, and human health. But the cost does not normally fall on the person who burns the high-sulphur fuel; only a part of the full social costs are private costs and influence private decisions.

Many of these externalities are caused by the segmented nature of government bureaucracies. For example, a highway department's mission is defined by statute. As the Department goes about its task of building the maximum amount of highway at the lowest possible cost, the California Department of Highways rips up neighborhoods, cuts through five-hundred-year-old redwood groves, and creates enormous social costs that never get into the department's cost-benefit calculations. Another example is the New York Sanitation Department (NYSD) which, when told to dispose of garbage, tows it offshore and dumps it. When the refuse washes back upon the beaches and estuaries of Long Island, the NYSD passes the blame to some other department.

External costs can come about in some exotic ways; for example, with products that ultimately interfere with some technique of protecting the environment. Polyvinyl plastic bottles produce no problems unless they are burned in a trash incinerator that is equipped with a scrubber designed to catch fly ash. The burning PVC plastic causes hydrochloric acid to form in the scrubber, destroying its metal casing. As technology produces more esoteric compounds, increasing attention must be given to such ultimate externalities.

A type of economic externality also may occur where resources are offered for sale in an imperfect market. The entrepreneur-owner of virgin land can realize a return from his investment either by marketing the aesthetic features of the land and its associate biota as a recreational and/or scientific resource, or by marketing it for its mining and logging potential. A rational, profit-maximizing entrepreneur will choose the alternative which yields the highest dollar returns.

The returns to our entrepreneur for the resource yield of his land result from interactions in the competitive market, and are priced as to their value in use by the highest bidder. The returns from the use of the land for recreational or scientific purposes result from the sale of a rare commodity which has no close substitutes, no alternative source of supply, and will be used in large part by consumers other than the bidder. Such a commodity will be valued in a noncompetitive market at that price per unit that the marginal buyer attaches to the commodity. Thus, the returns offered to the entrepreneur from the two alternatives are not comparable indices of social worth. The true social bid for the recreational or scientific resource would be the sum of the maximum that each individual consumer would have been willing to pay rather than go without the resource. The difference between this amount (that would have been bid in a competitive market), and the actual amount bid by a marginal buyer, reflects in some sense the external cost implicit in an imperfect market mechanism.

A similar argument can be offered that society is willing to bid little or nothing in an imperfect market to protect species threatened with extinction, or an entire ecosystem essential to the survival of a species. The instability that results from progressive reduction of biological diversity through monoculture presents a real social cost—one again external to the competitive market system.[7]

External effects are not unique either to environmental protection or to economics. The whole legal structure of society provides a mechanism for dealing with the "costs" that individuals and enterprises impose on others by their actions. Some of these legal devices, which we shall discuss in more detail in Chapter Six, include contract damages, tort liability, and criminal sanctions all of which are means of forcing decision-makers to "internalize" the external costs that their activities cause others to bear.

Without economic and legal sanctions, a decision-maker has no natural incentive to spend money to reduce harmful externalities. The fertilizer processor who dumps his chemical waste upstream from an adjacent fish hatchery rather than spend money for pollution control becomes wealthy and clean, while his fishery neighbor becomes impoverished and unhealthy.

If the damages caused by a decision-maker's externalities are measurable, and the external effects all fall upon one person or upon identifiable groups of people or organizations, and the source of the effects is known, then existing legal remedies are applicable.[8] Our fishery operator can sue in a court of law or equity, collect for past damages, and either collect liquidated damages as the present value of all future damages or seek a court order requiring the fertilizer processor to cease and desist from his dumping of chemical waste. If the damages involved are less than the cost of pollution control (and less than the net value of fertilizer output), the processor probably will choose to pay liquidated damages and will continue to dump his chemical waste. The economic effect of this arrangement is the same as if the adverse side effects had fallen upon the decision-maker in the first place—if the same person owned the fertilizer plant and the fishery, for example. The payment of damages causes just the right quantity of resources to be devoted to pollution control, even though the effects of the pollution are external to the decision-maker.

We may also find a voluntary contractual arrangement between the polluter and the victim which produces efficient pollution control, but which does not rectify the social injustice inflicted by the polluter upon his victim. Rather than endure the killing of his fingerlings by chemical waste in the water, our hatchery operator may contract to pay the fertilizer processor a sum of money in return for a certain amount of pollution control. This kind of situation will occur if the value of the reduction in pollution in fingerling longevity exceeds the amount of the payment, and if it costs less to control the pollution at its source than at the water intake of the hatchery. The payment insures that pollution will be controlled (but not necessarily eliminated) if the values gained or preserved are greater than the costs involved. Social justice is not served by such an arrangement, however, because whether the hatchery operator decides to suffer the effects of pollution, provide his own water treatment, or pay the fertilizer manufacturer to treat his wastes, the victim bears the cost of the external effects and the polluter enjoys a financial benefit to which he is not entitled.

Actual conditions obviously vary from the postulate of one polluter and one water-user. However, a multiplicity of users, the fact that the number of users will tend to increase over time and the changing technology of pollution controls does not alter the validity of the principles that arise from the two firm discussion. These factors do make it more difficult to apply those principles with much precision, to determine liquidated damages, or to calculate appropriate payments from the victim to the polluter to reduce his pollution. It is difficult to measure pollution from individual sources where there are a number of small-scale polluters; it is difficult to calculate the downstream benefits each water-user would derive from various combinations of waste water treatment by polluters. It is probably not possible to get all members of a diverse group of water- or air-users to bring joint legal action or to make joint payments for water or air quality improvement. However, neither social fairness nor economic efficiency are served if only a few of the victims of pollution extract damages from or make payments to only a few of the decision-makers responsible for pollution.

104. Cost-Benefit Analysis

Another recurring topic in the economic literature on environmental protection is cost-benefit analysis. Cost-benefit analysis is a practical way of assessing the long-term desirability of a project (in terms of looking at repercussions in the long run as well as in the short run), and its desirability from a broad view (in the sense of including the side-effects of pollution on all kinds of persons, industries, geographic regions, etc.) That is, cost-benefit analysis implies the enumeration and evaluation of all the relevant costs and benefits associated with a project. This analysis may involve a variety of the traditional areas of economics, including welfare economics, public finance, and resource economics, and may involve trying to meld these components into a coherent whole.[9]

Like any other investment, expenditures for environmental protection can be evaluated by cost-benefit analysis. Where capital outlays are required, the basic procedure is to forecast, for each year of the project's life, the probable benefits (usually damages which are avoided), less the operating costs, including depreciation. Since a dollar earned next year is worth less than a dollar presently in hand, future returns are discounted to their present value, with the discount factor based upon the estimated "opportunity cost" of capital—i.e., what the money could earn in its next best use.

If the total of discounted net benefits exceeds the capital cost being contemplated, then the project is economically reasonable and is put on a pile with other feasible projects to be ranked according to some other criteria—sometimes the ratio of benefit to cost, or sometimes on purely political considerations. If discounted benefits are less than anticipated costs, the project should be rejected on economic grounds, because a dollar spent on pollution control would be of less benefit than a dollar spent on damage abatement. That is, the marginal costs of environmental protection would exceed the marginal benefits.

The problem with cost-benefit studies is that those who have to pay for environmental improvement, notably businesses and municipalities, tend to inflate costs and deflate benefits, while those who have a vested interest in seeing the improvements carried out, notably environmentalists and involved agencies such as the Army Corps of Engineers, can be counted on to inflate benefits and minimize costs in their projections.

Some of the environmental degradation perpetrated by the Tennessee Valley Authority and others is carried out through the sophistry of cost-benefit analysis. One example was the Corps $15.3 million Gillham Dam across the Cossatot River in Arkansas, which was halted by an Environmental Defense Fund (EDF) sponsored injunction against the Corps. Three quarters of the benefits claimed for the Gillham, $970,000 annually, were in flood damage that the Corps said the dam would prevent. Yet on the 50 miles of flood plain below the dam there was virtually nothing to protect— in sum, three old wooden bridges, a dozen summer homes, and about 20 miles of gravel road. There had never been a recorded flood death on the Cossatot.

Upon inspection it appears that the figure of $970,000 was arrived at by the Corps through some circuitous reasoning. The dam was expected to result in a considerable growth in population and industry, which would mean that new buildings would be built on the present flood plain. It was the value of these *anticipated* structures that was being protected from flood by the dam. The only real beneficiaries of the dam, it turned out, were landowners who would reap a windfall profit as their forests were converted to industrial parks.

Another example was the proposed Tennessee-Tombigbee Waterway Project that the U.S. Army Corps of Engineers wanted to build to join the Tennessee River to the Tombigbee River in order to open up more of middle-America to the Gulf of Mexico and to foreign markets. The Corps claimed the Tenn-Tom project would produce $641 million in various benefits between the years 1980 and 2030 at a cost of $385 million for construction and $2.7 million per year in operating and maintenance costs. The Environmental Defense Fund won an injunction against the project in July of 1971 on the grounds that the benefits were overstated, the costs minimized, and the expected return on each invested dollar less than 10 cents.

Part of the dispute arose because the claimed benefits were all nonmonetary items, like time saved by shippers and increased recreation days, whereas the costs of the materials that were to go into the canal were very real monetary items. Paul Roberts, Economics Professor of the University of Florida, carefully evaluated the $641 million in claimed benefits on the Tenn-Tom project and concluded that a more realistic projection of the benefits would be approximately $17.5 million. The Corps, of course, disputed his figures.

Another element in dispute was the extremely low discount rate used by the Corps to arrive at the benefits. A low discount rate has the effect of maintaining the value of the claimed benefits at a very high level over the 50 to 100 year lifetime of the project, while the actual costs are written off within the first ten years. The discount rate for the Tenn-Tom canal was set at 3¼ percent, despite the fact that the Federal Water Resources Council was using 5⅜ percent for comparable projects elsewhere in the United States. Professor Roberts again suggested a discount rate of 9⅜ percent, which included the 5⅜ percent of the Council plus 4 percent for the element of risk involved in joining together two different ecosystems. Using the higher discount rate and the lower value of benefits, Roberts calculated a cost-benefit ratio of about 9 cents for the Tennessee-Tombigbee system.

Sometimes cost levels are overstated because the analyst considers only those solutions familiar to him and fails to include the full range of technological options available. Thus, in abating water pollution, most engineers think first of constructing plants for secondary or tertiary treatment of effluents. But in some parts of the country it would be cheaper simply to pipe untreated water to storage lagoons for settling, then use the waste water for irrigation and the settled solid for fertilizer or as landfill.

This type of option is already being used in Israel and the Soviet Union. Other techniques such as artificial aeration of streams, augmentation of low stream flow by planned releases from reservoirs, and storage of liquid wastes for discharge during periods of maximum stream flow may also be feasible.

Similarly, benefit levels may be understated because the analyst considers only physical damages (which is a difficult enough problem, because the relationship between individual pollutants and resultant damage can be both complex and highly variable). However, the principal benefit from pollution abatement may lie in reclaiming recreation areas, and in increased human aesthetic values that are difficult to measure and almost impossible to value monetarily. The problems involved in evaluating pollution cost in terms of damage to health and decrease in residential property values are discussed in the next chapter. It is probably fair to conclude that given the various biases to overstate and understate costs and benefits, the environmentalist is correct in arguing that most environmental cost-benefit analyses are biased in favor of rejection, in part because costs are overstated but principally because benefit levels are understated. However, this is far from a universal tendency, as is indicated by the Cossatot and Tenn-Tom examples.

The most glaring deficiency in these and similar studies is that no cost-benefit analysis is ever made on the alternative of leaving the Cossatot and Tennessee-Tombigbee rivers alone. Dam builders are in the business of building dams; if they seriously investigated alternatives to dams they might be investigating themselves out of a job. Robert Haveman of the University of Wisconsin, a resource economist with extensive experience in river development studies, estimates that only about 25 percent of the dams built by the Corps of Engineers would be justified if subjected to rigorous economic analysis, and that half of this 25 percent would be eliminated if environmental damages could be added to the cost. As of 1971, the Corps had built 255 dams in the United States, and 73 more were planned or being constructed.

105. The Myth of Unmeetable Costs

With President Nixon's Council on Environmental Quality claiming that the ultimate cost of reducing pollution to acceptable levels might run to $105 billion, industries and municipalities tend to think of all solutions as being virtually unattainable.[10] If one thinks of $105 billion as an instantaneous expenditure, this is virtually unattainable. But the total figure is misleading. Pollution abatement facilities are capital investments in equipment and facilities that take years to plan and install, and which have a long service life. On the basis of a 20 year amortization period, the $105 billion works out to an annual cost of $5.7 billion a year by 1975, about 1.6 percent of the projected gross national product and $28 per capita, which is one-sixth of the recent per capita expenditure of putting an American on the moon.

Of the total cost up to 1975, $62 billion would be for air and water controls, of which $35.6 billion would come from the private sector and the balance from federal, state, and local governments. All this supposes that industry is prepared to pay the $35.6 billion to meet 1975 standards. However, current figures for industry as a whole are not encouraging—industry's total planned 1972 expenditures for pollution control were only $3.84 billion; at 1972 rates it would take more than nine years to reach planned 1975 standards.

The figures for individual industries are more promising: the petroleum industry currently spends $360 million a year on pollution control and research. Based on sales of about $63 billion a year, pollution control amounts to less than one-half of one percent of industry's income. The inorganic chemical industry spends $414 million a year for waste treatment, which is 3 percent of its 1970 sales gross of $13 billion. The steel industry spends $320 million a year for pollution control, which is 2 percent of its gross industry sales. To reach its anticipated expenditure level of $2.64 billion by 1975, the steel industry would have to double its expenditures to a level of 4 percent of gross industry sales which would at most lead to a 2 percent price increase.[11]

The greatest expense comes from adding pollution controls to existing facilities. In new plant construction pollution control installations seldom are prohibitively expensive. The average for water pollution controls is under 10 percent of capital costs; for air pollution controls, under 5 percent. Older plants may have to have their conversion directly subsidized with government funds, or through invisible rebates in the form of tax concessions. Those older facilities that are operating so close to zero profit that pollution controls would bankrupt them are, by definition, surviving competitively by polluting. Arguments for using public funds to assist such plants must be based on equity and/or job preservation rather than on any economic rationale.

Some of the required money for pollution control facilities will come from tax-exempt bond financing. When Congress passed the 1968 Revenue and Expenditure Control Act that effectively eliminated most tax-exempt industrial bond financing, it made an exception for bonds that local governments sold to finance air and water pollution control facilities. Thus, Missoula County in the Big Sky country of Montana sold in 1970 a $15 million tax-exempt bond issue to pay for a special pulp mill furnace, an effluent clarifier to help abate water pollution, and equipment to modify a kraft paper production machine. Missoula County is leasing these facilities to the Hoerner Waldorf Corporation, a St. Paul paper company that operates four mills in the County. According to W. James Lopp, Vice President of Eastman Dillion, Union Securities & Company and head of its environmental finance unit, tax-exempt bond financing to pay for antipollution facilities would reach $300 million in 1972 and $1 billion in 1973. These estimates do not include pollution control facilities for electric utilities, which are the biggest single source of air and thermal pollution. If tax-exempt bond financing is also used by utilities, the total would be much higher.

The chief advantage of such borrowing to the corporation is the saving in interest costs. Over the 13 years from 1957 to 1970, corporate bond interest rates have averaged 1.50 percentage points, higher than similarly rated tax-exempt bond issues. During 1971 this gap widened to more than 2 percentage points. Using the 1.5 point spread, a tax exempt financing would result in a saving of approximately $3 million in interest costs over the life of a 20-year, $10 million bond issue. The second big advantage is that, for tax purposes, a corporation may deduct depreciation on pollution control facilities on an accelerated basis just as if it owned them outright.

106. Economic Side Effects From Pollution Control

There are some negative economic side effects arising from pollution control which are mentioned but not always spelled out in the literature. For example, pollution control standards of regional or national scope may prevent low-income or under-industrialized areas that are well equipped with water resources or air drainage flow from using these resources to attract new industries. Forbidding residents of low-income areas from trading a reduction in the quality of their environment for industrial employment may remove the only basis by which they can compete with more developed areas. But to permit low-income areas to so compete for industry is to reduce the incentive for industrialized areas to enforce pollution control standards and risk the loss of existing or potential new industry to less pollution control concerned regions of the country. Such industry-location considerations may be significant. Water-using industries accounted for 41.2 percent of manufacturing employment and 49.0 percent of manufacturing payrolls in the United States in 1967.[12] There is thus a large sector of the economy in which interregional competitive effects of water pollution control are likely to occur.

The application of uniform environmental standards to all geographic regions has another economic impact. A program of requiring all industrial water-using plants to meet the same waste treatment standards at their own expense penalizes older urban areas whose industrial facilities were neither located nor designed for efficient waste-water treatment, and improves the competitive position of areas where good sites for water-using manufacturing facilities are plentiful.

Incentives for adoption of pollution control standards may provide either a positive or negative impetus for the development of more effective waste treatment technology. For example, a strong federal subsidy, or tax concession program to encourage construction of sewage treatment facilities produces heavy short-term investment in facilities utilizing existing technology. Where such incentives are available for only a short period of time, or where there is fear that they will be withdrawn in the future when the problem becomes less current, investment in research to improve methods of waste treatment will be discouraged. Conversely, where it is likely that subsidies for waste treatment facilities will be available in the future, research on improved technology will be stimulated; but municipalities and industrial firms will postpone any construction of facilities that they would otherwise have carried out at their own expense.

107. Effects on Gross National Product

Econometric models of the United States economy have been used to estimate the effects of an altered investment mix resulting from the imposition of additional pollution controls. Economist Robert Anderson chose one such model that had accurately predicted changes in the economy during the 1962-64 period and added to it the assumptions that manufacturing industries were required to increase their investment on air pollution control at an annual rate of $1.2 billion; that public utilities were required to increase their rate to $320 million; and that new car prices rose by 1 percent due to the installation of emission control devices. Anderson then reran the model for the 1962-64 period.

At the end of the test period, gross national product (GNP) was down, at an annual rate of $617 billion, whereas without the assumed pollution controls the model had predicted more than $625 billion. Unemployment was up to 5.3 percent instead of 4.8, and the price level was 1.2 percentage points higher than it had been without pollution controls.

The result of Anderson's analysis is suspect for several reasons, and generally misleading in any case. An econometric model such as that used in the Anderson experiment is biased towards pessimistic predictions, in part because it does not (although it could) take into account the likelihood of improving the technology of pollution control. If we were to stop treating the use of the environment as a free good, pollution control would become an integral (and less expensive) part of production processes rather than an add-on; a great deal more research money and attention would be allocated to pollution control; and the cost of such control would almost certainly decline sharply.

Gross national product might actually increase in the intermediate run after the imposition of compulsory pollution abatement. A forced recycling of waste has proven for many companies to be much more profitable than waste "disposal" once they got involved in it. Also, pollution abatement makes urban areas more desirable, and people who work in desirable places do not have to be paid as much as those who work in undesirable places. A 3 percent decrease in the wages required to attract workers to New York City would more than offset the total cost of pollution abatement there. If a decrease in air and noise pollution were to reduce employee absenteeism or turnover, or were to increase productivity even marginally, costs of abatement could be similarly absorbed.

Actually, the real measure of economic and social welfare is not income at all, but the condition of the person or of society—which has only a slight relationship to GNP. Purer water for recreation may enrich leisure time satisfactions for many people, but qualitative improvement is not reflected in the gross national product. A decline in air or noise pollution may increase property values, but this is reflected in GNP only to the extent that rents increase. When somebody pollutes something and someone else cleans it up, the cleanup is added to the GNP but the pollution is not subtracted. Gross national product could actually decrease while human satisfactions increased, because GNP as currently calculated includes a large component of negative goods and services resulting in part from various pollutions.[13]

108. Social and Economic Systems

Pollution control is of concern to all the nations of the world, but especially to the technologically advanced or overpopulated ones. As the following survey indicates, approaches to pollution abatement have been no more successful in socialistic than in capitalistic economies; no more successful in eastern than in western societies to date.

Japan. Paradoxically the Japanese, with a culture strongly based on the harmony between man and nature, have handled the problem of environmental pollution worse than almost any other society in the world. Japan's economy, with a unique blend of private enterprise and governmental central planning, has forged ahead in every area to achieve a gross national product that ranks third behind the United States and the Soviet Union. Unfortunately the price that has been paid for 25 years of industrial development has been such extensive damage to the Japanese environment that the most heavily populated areas from Tokyo to Kobe to Hiroshima threaten to become unlivable. To begin to reverse the downward environmental spiral would entail a virtual rebuilding of the country. For example, in most Japanese cities, the infrastructure basics such as sewage systems are either totally inadequate or absent. And despite the widespread ownership of automobiles, Japanese roads are so inadequate that the world's largest and densest traffic jams occur in Japanese cities, adding their share of air and noise pollution.

Europe. The pollution of beaches, rivers, lakes, fjords, and seas was the subject of debate in 1970 in countries as diverse as Norway, West Germany, Switzerland, and Spain. A *World Life* spokesman quoted by *The New York Times* estimated that the United States was spending 25 times as much per capita fighting pollution as were Britain and France. Perhaps the worst problems are in the Mediterranean area, where ecologists are seriously concerned that the Sea itself may be slowly dying. The Mediterranean is fed by the great rivers of the Continent and all these rivers, in particular the Italian ones, are grossly polluted. The situation is critical because many of the chemicals carried to the Mediterranean from Italian plants contain substances which cannot be broken down biologically and which collect in a thin layer on the surface of the water. These chemicals are absorbed by simple plant and aquatic life, travel up the food chain and accumulate in fatty tissues of fish.[14]

Even Ireland, which was a late starter in the industrial revolution, is having its problems. Untreated sewage has been pumped into the lakes of Killarney until it is not now safe to swim in Sandymount Strand because Dublin Bay is considered an "area of high pollution." The stench from the lower reaches of the River Liffey when the tide is low and the wind is from the east is overpowering. There is speculation that in 10 years no fish will be able to survive in Lough Gill, County Sligo, because of effluents from a nylon factory which is being built on the shores of this lake.

There are some bright spots, however. One of the more significant environmental victories occurred in England in April, 1971, when the British Government ruled out an inland site for the new London airport. The alternative chosen was Foulness, a sparsely populated area on the east coast of England, 55 miles from London. Foulness is farther from London than any of the inland sites considered; the cost of locating an airport there with access facilities was estimated at $1.2 billion, which is $360 million more than the cost of the next alternative site.

This decision reversed the recommendation of a high-level commission headed by a High Court judge, Sir Eustace Roskill, which suggested a site north of London near Cublington in the Vale of Aylesbury in Buckinghamshire. The Cublington airport would have obliterated a good deal of rolling countryside, and the village of Stewkley, location of a 12th century Norman church which would have been demolished to make room for a runway.

The Soviet Union. The U.S.S.R. (which introduced central planning fifty years ago), also has disfigured its countryside with strip mines, industrial pollutants, and overfarming. Its apartment and office construction programs have produced in Moscow perhaps the most aesthetically depressing cityscape in the world. Given the environmental errors of the United States and western Europe as examples, the Soviets have introduced no innovations in city or highway planning, in waste disposal, or in automobile design or highway construction. Replacing private property with public ownership has produced a level of environmental disruption almost as extensive as that in the United States. According to Marshall I. Goldman:

Comparing pollution in the United States and in the USSR is something like a game. Any depressing story that can be told about an incident in the United States can be matched by a horror story from the USSR. For example, there have been hundreds of fish-kill incidents in both countries. Rivers and lakes from Maine to California have had such incidents. In the USSR, effluent from the Chernorechensk Chemical Plant near Dzerzhinsk killed almost all the fish life in the Oka River in 1965 because of uncontrolled dumping. Factories along major rivers such as the Volga, Ob, Yenesei, Ural, and Northern Dvina have committed similar of-

fenses . . . there is not one river in the Ukraine whose natural state
has been preserved . . . in Sverdlovsk in 1965, a careless smoker
threw his cigarette into the Iset River and, like the Cuyahoga in
Cleveland, the Iset caught fire.[15]

By far the best known example of the misuse of water resources
in the USSR is the condition of Lake Baikal, a four hundred mile
long body of water in southeastern Siberia which is thought to be
the oldest body of water on earth. With a maximum depth of more
than 5000 feet, it is also the deepest and largest volume body of
fresh water on earth. Since much of the Lake Baikal watershed area
is surfaced with rock, the water is exceedingly pure. There are over
1200 species of living organisms in the lake, including the *nerpa*,
the world's only fresh water seal, and 700 other organisms that are
found in few or no other places in the world. The deterioration of
the lake's water quality has been chronicled by Marshall Goldman:

In 1966, first one and then another paper and pulp mill appeared
on Lake Baikal's shores. Immediately limnologists and conservationists protested this assault on an international treasure. None-
theless, new homes were built in the vicinity of the paper and
pulp mills, and the plant at the nearby town of Baikalsk began to
dump 60 million cubic meters of effluent a year into the lake. A
specially designed treatment plant had been erected in the hope
that it would maintain the purity of the lake. Given the unique
quality of the water, however, it soon became apparent that almost
no treatment plant would be good enough. A few months after this
effluent had been discharged into the lake, the Limnological Institute reported that animal and plant life had decreased by one-
third to one-half in the zone where the sewage was being discharged.

Several limnologists have argued that the only effective way to
prevent the mill's effluent from damaging the lake is to keep it out
of the lake entirely. They suggest that this can be done if a 67
kilometer sewage conduit is built over the mountains to the Irkut
River, which does not flow into the lake. So far the Ministry of
Paper and Pulp Industries has strongly opposed this, since it would
cost close to $40 million to build such a bypass. They argue that
they have already spent a large sum on preventing pollution. Part
of their lack of enthusiasm for any further change may also be explained by the fact that they have only had to pay fines of $55 for
each violation. It has been cheaper to pay the fines than to worry
about a substantial cleanup operation.

There are those who see even more dire consequences from the exploitation of the timber around the lake. The Gobi Desert is just over the border in Mongolia. The cutting of the trees and the intrusion of machinery into the wooded areas has destroyed an important soil stabilizer. Many scientists report that the dunes have already started to move, and some fear that the Gobi Desert will sweep into Siberia and destroy the taiga and the Lake.[16]

The Soviet government announced in early 1971 that it was going ahead with another cluster of wood pulp mills on the shore of Lake Baikal in the town of Selenginsk near Kamensk, about 100 miles northeast of the first mills at Baikalsk. The announcement said that "An advanced three-state treatment system will remove all toxic material from waste waters and preserve Baikal as one of the cleanest lakes of the world." Officials of the pulp and paper industry contend that the forests in the Baikal region must be cut because the Soviet Union badly needs pulp for the cord used in automobile and aircraft tires.

Many of the explanations for environmental pollution under capitalist systems also hold for state-owned, centrally-planned economies such as the Soviet Union. The Russians have also been unable or unwilling to consider external diseconomies in their accounting systems. A Russian plant manager is charged for labor, raw materials, and capital equipment, but not for the social costs of production arising from air, water, noise, and other pollutions. As in capitalist countries, air and water are treated by the Russians as "free" goods. The Russians also have to contend with the same population growth pressures, the same rural-urban shift, and the same pressures from expanding technology as the West.

There may also be special incentives to pollute built into the Soviet economic system. Russian decision-makers, both managers and state officials, are judged first by the rate at which economic production is increased, and second by the efficiency (the ratio of inputs to outputs) with which this is accompanied. Such officials could be expected to identify with polluters rather than conservationists; polluters increase economic growth, while conservationists divert resources away from increased production. In their approach to economic expansion, Soviet decision-makers are not unlike officials in states such as South Carolina, who have successfully induced chemical and other industries to relocate by the promise of lax or nonexistent environmental protection requirements.

As Goldman has pointed out, there are some things the Russians do very well:

> The Russians have the power to prevent the production of various products. Thus, the Soviet Union is the only country in the world that does not put ethyl lead in most of the gasoline it produces . . . The Russians have not permitted as much emphasis on consumer-goods production as we have in the West. Consequently there is less waste to discard. In the USSR there are no disposable bottles or diapers to worry about. It also happens that, because labor costs are low relative to the price of goods, more emphasis is placed on prolonging the life of various products. It is worthwhile to use labor to pick up bottles and collect junk. No one would intentionally abandon his car on a Moscow street, as 50,000 people did in New York City in 1969. Even if a Russian car is 20 years old, it is still valuable.[17]

The Soviet experience indicates that while in theory a centralized and totally planned economy could have advantages over other economic systems in solving environmental disruption, theirs has not to date done so to any appreciable extent.

109. International Pollution Agreements

Given the universality of pollution problems among developed nations, there is an emerging need to prevent pollution-control costs in one country from becoming a competitive disadvantage in trade with other countries. A more rapid pace of environmental improvement in the United States, without consideration of costs, could price American exports out of world markets for products that are "pollutant heavy," such as steel and automobiles. Varying national pollution standards could also become a barrier or incentive to foreign investments, making some foreign countries pollution havens for low cost production.

One result could be increased protectionist demand for import restrictions against countries which are lax in enforcing pollution standards—something that multinational companies vigorously oppose. A more promising solution would be cooperative international action towards agreement on detailed standards that would keep pollution costs from disrupting international trade. Proposals for such cooperative actions were discussed at the United Nations Conference on the Human Environment in Stockholm in

June, 1972 but no agreement was reached (see Chapter Eight). An agreement could be modelled after the General Agreement on Tariffs and Trade (GATT), which has set basic principles of fair trading practices and has led to the reduction of tariffs and other trade barriers.

A problem might arise with countries like Japan which believe that government subsidies should be given to industry for pollution control. Such subsidies produce distortions of costs as compared to countries where manufacturers are required to pay their own pollution control costs. Also, differing production techniques present an obstacle to any standards agreement. For example, the American steel industry still relies basically on the open-hearth production method, while many other countries use the oxygen method, which produces two to three times more pollutant per ton of steel.

Any agreement on standards would initially include only the industrialized countries; economically and morally, the less developed countries could not be forced to meet the same initial standards. However, negotiating a date when such standards would have to be met, and the sanctions to be imposed if they were not met, might prove to be a messy political problem.

DEFINITIONS AND CLASSIFICATIONS

Pollution, therefore, is a worldwide problem, and so are its costs. Before discussing specific kinds of pollution in the following chapters, however, we need to define some of the concepts that will be used, for example, what is a pollutant? what is waste? what are the components of the "cost" of environmental protection? In part these definitions are necessary for clarity, in part because the way we define things strongly influences the way we think about them. The remaining sections of this chapter define a number of terms, then classify pollutants according to their physical, chemical, or behavioral characteristics, and finally discuss the important distinction between normative pollution levels and optimal pollution levels.

110. Some Definitions

Pollutant. A pollutant, in a very general sense, is anything animate or inanimate that by its excess reduces the quality of living. Thus, sulphur dioxide is a pollutant, so is excessive noise, and so is an epidemic of rats.

One idea that is important in understanding pollutants and pollution is that, while man is unquestionably a polluter, man is not (in a pollution context), a consumer. "Consumer" is a word invented by schools of business and by Madison Avenue, and has no real application to a discussion of pollution. Man consumes nothing, whether it be the food he eats, the automobiles he drives, or the clothes he wears. He merely "uses" things, and, according to the law of Conservation of Matter, discards exactly the same mass of material after use. Sometimes, as in the case of a building or a dam, disposal is postponed for a very long time. Much of the weight of each year's production is transformed into gas and released into the atmosphere without any special treatment. But the concept of nonconsumption still holds, and becomes especially important when we begin to think in terms of reconversion.

Directly related to the concept of nonconsumption is the fact that there is not really an air pollution problem or a water pollution problem, but rather a materials disposal problem. To eliminate air and water pollution may simply mean transforming them into the problem of disposing of solid waste, another pollution problem.

Waste. Waste, the component of most pollution, is defined as some substance that our society does not as yet have the intelligence to use, or that our present economic system prevents us from using or reusing. Waste disposal—which is what people do with waste—is not capable of economic definition because it is really a misnomer. Disposal is usually used to denote the process of burning, diluting, grinding up, or spreading things around so that they won't be offensive or be noticed. It is a little like the lazy housewife who, when asked what she did with her garbage, said "I just kick it around until it gets lost."

Disposal. Disposal is really just a conversion process. When we convert iron ore to iron, we call it smelting or refining; when we convert the iron into a car, we call it manufacturing. When we stop using the car and abandon it, we call it waste disposal. In the latter case we have converted the automobile into a large chunk of iron oxide. We could as well, if not as easily, have melted the car and converted it back into iron.

If our present technology does not allow us to reuse iron oxide,[18] an alternative method of disposal might be to take these chunks of metal from compressed automobiles to the mid-Western plains, to Kansas or Saskatchewan, and use them as building blocks for the construction of mountains which could then be used for skiing and other recreation. These would be mountains of fairly pure iron, which could be mined when our economics changed or iron ore became scarce.

In the same sense the use of water is merely a conversion from clean water to dirty water. The reconversion of water would be the cleaning of it. In economic terms there is no producer and no consumer of water, only consecutive steps in the conversion process.

All this suggests a shift in emphasis from the temporary expedient of waste disposal to the more permanent solution of waste recycling and reconversion. However, if producers manufacture synthetic materials that are not degradable (reconvertable) by nature, the conversion cycle is broken; after use, someone must unmake these things using the same technology. This implies that before designing things and converting resources to new forms, producers should have in mind a plan for reconversion and recycling. Some nerve gases, for example, can be broken down into nonpoisonous components; some equally potent gases cannot be safely converted, and must be disposed of in the ocean or similar places.

Some kinds of reconversion are hard to envision; for example, collecting old newspapers, cleaning the print off them, and reusing them for something else. Experiments are taking place; for example, in Beltsville, Maryland, where cattle are being fed pure newsprint cellulose with amino acids and other additives, which the ruminants change into protein. One may visualize a factory where old newspapers go in one end and steaks come out the other. That is the sort of objective to which recycling is striving!

Landfill is another form of reconversion or recycling, in this case again using wastes as a construction material. Untreated waste (garbage, sewage sludge, building rubble) is buried in layers, each covered by several inches of compacted earth. This technique has transformed thousands of acres of low-value land into parks, playgrounds, golf courses, and other useful facilities.

Reconversion and recycling would be helped by a tax on waste (for example on throw-away containers) imposed on the basis of the difficulty of recycling. There always seems to be some perverse component that gets in the way of economic recycling; the aluminum ring that a twist-off cap leaves around the neck of a soft drink bottle makes it uneconomic to grind such bottles into cullet for glassmaking, and the cost of removing the metal, either before or after grinding, is prohibitive. Similarly, the tin coating and lead solder on the otherwise steel "tin can" largely exclude it from being economically recycleable. The government could also help through discriminatory purchasing policies, as federal government purchases account for about 6 percent of all packaging expenditures. If government were to insist upon tin-free steel cans or aluminum-ring free soft drink packages, industry would have a powerful economic incentive to alter its technology.

Cost of Environmental Protection. To continue with definitions, the *cost* of environmental protection, to society, is the sum of expenses incurred to prevent environmental damage, plus expenses incurred through not preventing environmental damage. Thus:

Cost of environmental protection = expense of preventing environmental damage + expense of environmental damage not prevented.

The expense of preventing environmental damage is easily measured; it is the total expense incurred by public and private parties to prevent damage caused by waste products. Expenditures by public bodies for sewage treatment, and by private corporations to remove soot from smoke are good examples.

The expense incurred from environmental damage *not* prevented is much more difficult to identify and to measure. Conceptually, it is the expense of pollution; the money value of the damages caused by waste products after they are released into the environment.

The following are three components of such expenses:[19]

Expense of environmental damage not prevented = public expenditures to avoid pollution damage + private expenditures to avoid pollution damage + the money equivalent of the welfare cost of protection.

The first component of this equation is public expenditures to avoid pollution damages—not to prevent pollution, but to prevent the damage that pollution causes. For example, the cost of treating drinking water to prevent typhoid epidemics is such a public expenditure, while the expense of sewage treatment is an expenditure to prevent pollution.

The second component of the equation is expenditures made by private parties to avoid pollution damage. Examples are the expenditures that individuals who live in areas of air pollution make compared with those who live in cleaner-air areas for dry cleaning of clothes, painting of houses, home air-purification systems, and so on.

The third component of the equation is the money equivalent of the welfare cost of pollution—the dollar value of the reduction of public welfare from pollution damage that is not prevented. This includes, for example cost of foregone or lower earnings or of increased commuting expenses where families move away from city centers to escape air pollution. A manufacturer that moves to a new location to secure clean water, but that must then pay higher transportation costs for raw materials or finished goods, incurs such costs. Similarly, the value of pleasures foregone or diminished —the lack of local swimming because of polluted water, or the decreased aesthetic value of a littered landscape—are real losses in welfare that at least some people are willing to pay to have reduced.[20]

Given these components of cost, the problem facing the economist is to minimize the total cost of environmental protection; the cost of preventing environmental damage, plus the cost of environmental damage not prevented.

A final area of definition concerns the boundaries of a polluted area which are relevant for the study or control of the pollution problem at hand. Unfortunately, such boundaries are seldom if ever well defined. No area is completely isolated from other areas, since many pollutants are fairly mobile, and different pollutants affect areas of vastly different sizes. Airborne soot of a given size range may affect an area of from 20 to 400 square miles, while radioactive gases may contaminate the entire atmosphere. Different wastes discharged into a water system can affect either large or small areas before they are diluted or degraded below noxious concentrations. Areas of population concentration such as metropolitan areas would seem natural areas for the study of pollution, but this is not feasible where pollutants and pollution victims are mobile over greater distances, as is true with automobiles without emission control systems.

If an arbitrary choice of area must be made, most often it consists of choosing a political area, because pollution control is closely tied to political unit decisions. A political unit such as the State of Maine is clearly too large for some kinds of pollution control and too small for others; some pollution from the Canadian Maritime provinces affects people in Maine, while some pollution from Portland, Maine does not even extend to North Portland. But the residents of Maine do have some political clout in controlling those pollutants which arise within their state, and less clout (but perhaps some influence) in abating pollution which arises outside Maine's borders.

111. Classification of Pollutants

Now that we have defined the most common concepts to be used in the following analysis, it might be helpful to classify kinds of pollutants according to their physical, chemical, or biological characteristics, or according to their behavior when released into the environment. Various pollutants may be usefully classified in terms of their toxicity, the productivity of their source, their durability, avoidability, and the costs of abatement.[21]

Toxicity. Probably the most critical criterion for classifying pollutants is the toxicity of the pollutant to plant and animal life. However, there are wide differences of opinion on the short-term and cumulative effects of various dosages of toxic pollutants, for example, the extent to which toxic materials are concentrated by successive organisms in the food chain and by long-lived organisms such as man.

Productivity. A second basis for classifying pollutants is by the economic benefit which arises from the activity which causes the pollution. It is often said that a paper mill smells like money to those who live nearby. If most of those living near the mill depend on it for their livelihood and have no reasonable alternative sources of income; if temperature inversions occur only occasionally and dangers to health are minimal; then residents of the area may prefer the odor to the alternative of shutting down the source of pollution.

Durability. The durability criterion requires that the highest priority in pollution prevention be given to conditions requiring many years or decades to clean up, and that lower priority be given to pollution the effects of which can be remedied fairly quickly when efforts are made to do so. Thus, in the case of water pollution the highest priority would be assigned to controlling the pollution of large lakes in which the water is retained for decades, and a lower priority to the prevention of population of fast-flowing rivers the beds of which are flushed yearly by spring floods.

Avoidability. The economic damage caused by pollution varies with the ease with which people may avoid using the polluted resource. Thus, in an area with substantial ground water, numerous unpolluted streams and ample precipitation, the pollution of one particular stream does not impose large opportunity costs on prospective users. The same quantity of stream pollution would impose much higher costs if there were no readily available alternatives to the use of the water, either because other streams were already polluted, or because alternative streams did not exist.

Adoption of an avoidability criterion means that air pollution would be given greater priority than water pollution, since people are under no physiological or other compulsion to come into contact with water in any particular place (except for drinking and washing water, which is normally treated), while there is no way for people to avoid breathing polluted air if they live or work in an area where air pollution exists. By the criterion of avoidability, air pollution is to people what water pollution is to fish.

Cost of Pollution Abatement. Some industries have wastes that are much more expensive to treat than are those of other industries. Given a similar set of physical conditions, wastewater treatment plants for industrial use may be quite expensive compared with treatment plants for sewage. If the self-cleansing capacity of a waterway is considered as a productive natural resource to be used for the public benefit, that benefit is maximized if the resource is used to substitute for the most costly industrial and municipal waste treatment. A cost-of-pollution-abatement criterion implies that the public interest is better served by diverse water-treatment standards than by uniform ones that require a predetermined level of waste treatment by all industries and municipalities.

112. Normative and Optimal Pollution

A final concept of importance for following chapters is the distinction between normative and optimal pollution levels.

The chapters which follow are not concerned with any concept of a normative (or "ideal") environment, or with normative levels of pollution, since no idealized standards do or can exist. There are no definitions of normative environmental conditions that have any moral superiority over others, except by reference to the selfish needs of one portion of society over another. For example, is it good or bad for plants to alter the atmospheric composition in favor of oxygen, or for animals to do so in favor of carbon dioxide by breathing oxygen and by eating plants? Is it worse to kill stands of cedar by industrial fumes, than to cut cedar trees in order to build housing for the poor?

A number of pollutants are natural constituents of the air. Even without man's technology, plants, animals, and natural activities would cause some pollution—volcanic action would release sulphur dioxides, and surface winds would stir up particulate matter. There is no way to remove all pollution from the air. The "right" composition of the atmosphere is that which contains some oxygen, some carbon dioxide, and some hydrogen sulfide, in combinations which permit organized society to pursue the greatest possible satisfaction for its human members.[22]

The "correct" solution to our current environmental problems will not produce pure air, or pure water, but rather some optimal state of pollution. The cost of this optimal state is best expressed in terms of the other goods—such as more housing, more ballet, more medical care—that must be foregone in return for somewhat cleaner air and water. Society will be willing to forego one ballet company, or one housing development, only if the resources that would go into it yield less human satisfaction than would the same resources if they were devoted to the elimination of air or water pollution. Trade-off by trade-off, an economist would divert our productive resources from current goods and services to the production of a cleaner, more pastoral nation up to the point where society values the next ballet or house more highly than it values the next unit of environmental improvement that the diverted resources would create.[23]

REFERENCES

[1] Alvin Toffler, *Future Shock* (New York: Random House, 1970).

[2] There are economic incentives that might encourage a voluntary reduction in population growth. We might subsidize or pay in full the cost

of abortions and voluntary sterilizations, and seek ways to give women more equality in educational and job opportunities so as to aid them in developing professional interests to compete with family interests. Using the tax system, we might stop taxing single persons more heavily than married ones. We could grant a standard exemption for the first two children in a family, and thereafter levy a tax on an increasing scale. For those who do not pay a tax, we could give women between the ages of 17 and 45 who already have two children a cash bonus each year they do not get pregnant. Each man or woman who volunteers for sterilization could be given a similar cash bonus or tax credit.

[3]Barry Commoner, *Is There An Optimal Level of Population?*, a paper presented to the Annual Meeting of the American Association for the Advancement of Science, Boston, Mass. (December 29, 1969).

[4]Paul Ehrlich, *The Population Bomb* (New York: Ballantine Books, Inc., 1968), in particular pp. 15-67.

[5]Paul R. Ehrlich and Anne H. Ehrlich, *Population/Resources/Environment: Issues in Human Ecology* (San Francisco: W. H. Freeman, 1970), p. 300.

[6]The Olin Corporation, producer of 20 percent of the U.S. supply of DDT, announced on June 30, 1970 that it was going out of the DDT business. Olin's announcement came after a suit to halt the discharge of DDT from Olin's Redstone Arsenal plant into the Wheeler National Wildlife Reserve had been filed in U.S. District Court for the District of Columbia by the Environmental Defense Fund, National Audubon Society, and National Wildlife Federation. The suit alleged that the concentrations of DDT in the Olin discharge ditch at the point of discharge from the plant ran as high as 460 parts per billion, when 1 to 3 parts per trillion of DDT caused such accumulation in the fatty tissues of fresh water fish as to render them unfit for human consumption, and 2 to 3 parts per trillion caused reproductive failure among many species of carnivorous birds and fish in other ecologically comparable areas.

[7]F. Bator has suggested that there are three possible, but not mutually exclusive types of externalities that may lead to market failure: ownership externalities, technical externalities, and public goods externalities.

In the case of ownership externalities, the basic cause of market failure is the inability of the owner of a factor of production (because of legal or other reasons) to charge for the value of his services. Ownership externalities play an important part in the failure of the private market to provide a sufficient quality of water for outdoor recreation.

Technical externalities are due either to indivisibilities or to increasing returns to scale. Production may occur when price equals marginal cost and exceeds average cost, with the position representing a local profit maximum rather than a global profit maximizing point. Thus, there is failure by signal.

Public goods externalities occur when an individual's consumption of a good leads to no subtractions from any other individual's consumption of that good. Thus, there is no set of market prices for public goods which is useful for individual production and/or consumption decisions. In the absence of a set of market prices to ration any fixed supply of public goods, a private market for public goods will fail by existence.

Source: F. Bator, "The Anatomy of Market Failure," *Quarterly Journal of Economics*, No. 72 (August, 1958). An illustration of how a blend of the three types of externality conditions leads to market failure in the provision of water recreational facilities in an estuary is given in Paul Davidson, F. Gerard Adams, and Joseph Seneca, "The Social Value of Water Recreational Facilities Resulting from an Improvement in Water Quality: The Delaware Estuary," in Kneese and Bower, *Water Research* (Baltimore: Johns Hopkins Press, 1966), pp. 179-187.

[8]See Otto A. Davis and Andrew Whinston, "Externalities, Welfare, and The Theory of Games," *Journal of Political Economy*, No. 70 (June, 1962), pp. 241-262, for an excellent discussion of the measurement of externalities.

[9]The best overview of the topic known to the writer is A. R. Prest and R. Turvey, "Cost-Benefit Analysis: A Survey," *The Economic Journal* (December, 1965), pp. 683-731.

[10]Estimate in the Second Annual Report on the State of America's Environment by the Council on Environmental Quality (1971). The figure of $105 billion represents a doubling of the 1971 rate of spending on environmental protection from all sources by 1975. Of the total funds it is projected that 23 percent would go to air pollution control, 36 percent to water pollution control, and the rest to solid waste management.

[11]To polluting industries that plead poverty it might be pointed out that identical defenses were once raised on behalf of slavery, child labor, and the working conditions and housing associated with the early Industrial Revolution in England. The same defenses are offered for migrant farm workers camps in California and the southwest.

[12]U.S. Bureau of the Census, *Census of Manufacturers, 1967* (Washington, D.C.: U.S. Government Printing Office, 1968).

[13]For an overview on the fallacies of GNP see Kenneth E. Boulding, "Fun and Games With The Gross National Product—The Role of Misleading Indicators in Social Policy," in Harold W. Helfrich, ed., *The*

Environmental Crisis (New Haven, Conn.: Yale University Press, 1970), pp. 157-170.

[14]See John Cornwell, "Is The Mediterranean Dying?", *The New York Times Magazine* (February 21, 1971), pp. 24-25, 47-57.

[15]For a succession of U.S. type environmental horror stories in a Soviet setting, see Marshall I. Goldman, "The Convergence of Environmental Disruption," *Science*, No. 170 (October 2, 1970), pp. 37-41. The discussion which follows is based on this article, and on Marshall I. Goldman, "The Pollution of Lake Baikal," *The New Yorker* (June 19, 1971), pp. 58-66.

[16]Goldman, "Convergence," pp. 38-39.

[17]Goldman, "Convergence," pp. 41-42.

[18]In the remelting of automobile bodies, it is the small amount of copper in the wiring and the motors which pollutes the steel. If we were concerned about being able to convert car bodies back into iron we would design automobiles with aluminum harnesses and ferrite magnets which would oxidize and come off as slag so that we could get a better grade of steel.

[19]The classification and much of the analysis comes from J. H. Dale's *Pollution, Property and Prices* (Toronto: University of Toronto Press, 1968), pp. 12-26.

[20]The amount of information available on the third component is scant; our knowledge of how to collect such information is extremely limited. The average citizen has only the vaguest idea of what air pollution costs him in terms of lower earnings or increased commuting fees, to say nothing of the cost of excess cleaning, painting, or health bills. A statistician might try to collect such data by comparing per capita expenditures on these items in locations with different amounts of air pollution, but the data collection problems and costs are staggering. How would a statistician go about measuring the value of damage done to such intangibles as aesthetic enjoyment of natural environments located several hundred miles away, and which our citizen visits only at intervals of several years? How would he calculate the value of a day's sickness because of bronchial irritation caused by sulphur dioxide particles in the air? The respondent will want to know whether we are talking about a work day or a vacation day of illness; whether the malady is sufficient to require a day in bed, bed plus drugs, or a hospital visit plus curtailment of smoking for a month. And those are simple questions compared with the proper one of asking the respondent how much he would be willing to pay to prevent the sulphur dioxide content of the air in his city from exceeding, say, five parts per million more than ten days a year—especially when one is also unable to tell him speci-

fically what the effects of such a concentration on his personal health might be.

[21]The classification criteria are elaborated in R. Stepp and S. Macaulay, "The Pollution Problem," *Legislation and Special Analyses of the American Enterprise Institute for Public Policy Research*, 90th Congress, 2nd Sess. (No. 16, 1968), pp. 26-30.

[22]A thoughtful expansion of these ideas is contained in Lectures One, Eight, and Nine of the Harvard Alumni Summer College Lecture Series (1970), given by Leonard M. Ross.

[23]There is an economically optimal level of purity for a portion of the environment, say a river. Writers have argued that this level of purity will be obtained if the parties involved can bargain over the level of pollution to be tolerated. A willingness to pay for what one desires indicates the relative value of pollution or nonpollution to each of the parties; the one with the highest value will have his way. However, where *many* parties are affected by the actions of one or more persons, it may prove impossible to organize for bargaining purposes. This problem is discussed in more detail in Chapter Five. See also Ronald H. Coase, "The Problem of Social Cost," *Journal of Law and Economics* (October, 1960), pp. 6-15.

2

Air

Pollution

Baa, baa, black sheep,
What's dirtied up your wool?
Soot, smoke, sulphur
Of which the air's so full.
Soot from the foundries
Sulphur from the stacks,
And that's why we sheep now wear
Black coats on our backs.

200. The Extent of Air Pollution

Air pollution is not a recent world phenomenon. As
early as the beginning of the Industrial Revolution many communities endured levels of smoke pollution that would be considered
intolerable by present standards. In the last half of the 19th century, a number of citizen groups picketed the British Parliament
to protest the smoke-laden air of London; their protests were lost
in the desire of the government for industrial development at any
price. In the U.S., the cities of Chicago and Cincinnati passed
smoke control laws in 1881; by 1912, 23 American cities with
populations over 200,000 had passed similar laws, although they
were seldom enforced. By the 1930's and 1940's, smoke pollution
had become sufficiently bad in eastern and midwestern industrial
cities to cause improved smoke pollution legislation to be passed
and enforced, with control efforts primarily focused on cutting
down smoke from fossil fuels, particularly coal.

Although the number of recorded fatal air pollution incidents has
not been large, the precedent does exist. In 1930 in the Meuse
Valley in Belgium, 63 people died and more than 100 became
seriously ill from a combination of particulate and sulphide pollution. In 1948, in the Monongahela River Valley in Pennsylvania,

almost half the population of the town of Donora became ill and 17 died from air pollution which built up under a low-level temperature inversion. In London, in December of 1952, an estimated 4,000 excess deaths occurred during a two week period, followed by another incident in December of 1962 producing an estimated 300 excess deaths.

In the United States the present dimension of air pollution is severe. According to the United States Public Health Service, a total of 43 million persons in over 300 U.S. cities live under an air pollution hazard rated as major. An additional 30 million people live in 850 other cities with air pollution that is less than severe, but too serious to be classified as minor.[1] Testimony before a Senate committee in 1963 indicated that about 7,300 U.S. cities and towns, housing 60 percent of the population of the United States, had at that time a discernable air pollution problem of one kind or another.[2]

201. The Sources of Air Pollution

In terms of total national air pollution, the automobile is the greatest single contributor by weight. Nonautomotive sources of air pollution include emissions from the consumption of fossil fuel by industries such as pulp and paper, iron and steel, petroleum, refining and smelting, and chemicals; power plants which use fossil fuels to produce electricity; disposal of solid wastes by combustion; and the heating of homes, offices, and plants by fossil fuels.

Automotive emissions include carbon monoxide, hydrocarbons, oxides of nitrogen, lead compounds, sulfur dioxide, and particulate matter. It is estimated that 90 million automobiles and trucks annually discharge into the air 66 million tons of carbon monoxide; 6 million tons of nitrogen oxides; 12 million tons of hydrocarbons; a million tons of sulfur oxides; a million tons of particulate matter; and 190 thousand tons of lead compounds. The total from nonautomotive sources is 25 million tons of sulfur oxides, 11 million tons of hydrocarbons, and 5 million tons of carbon monoxide.

Industrial air pollution is not restricted to industrial centers of the United States. A twenty-four member combine called Western Energy Supply Transmission Associates (WEST) are in the process of building six of the largest fossil fuel power plants in the world in and around the Colorado Plateau, with the potential of turning the Four Corners area of Utah, New Mexico, Arizona and Colorado into a major pollution zone.

Already the single 2.1 million kilowatt plant at Farmington, New Mexico emits as much particulate matter per day—250 tons—as New York and Los Angeles combined. The plume of smoke and ash, which was the only man-made landmark that the Apollo 8 astronauts could consistently photograph from lunar orbit, drifts across more than 200 miles of Indian reservation. New Mexico ordered the company to make the stacks at Farmington 99.2 percent pure by the end of 1972, or shut down. The plant is installing wet scrubbers that will reduce the ash to 15 tons a day, but when all six plants are operating in the desert by the late 1970's, even if equipped with wet scrubbers, they will still emit 350 tons of fly ash a day. If not forced to clean up, they will also emit 2,160 tons of sulfur oxides a day—more than Chicago or New York City, and 850 tons (yes, tons) of nitrogen oxides per day. The nitrogen oxide figure rivals that of Los Angeles, which produces in the area of 900 tons a day from its millions of automobile exhausts.

Estimates of the total quantity of air pollutants released daily in the U.S. range up to 400,000 tons. While probably accurate, such figures are only relevant when compared with the total volume of air involved—a figure which to my knowledge has never been estimated. In the absence of such a comparison, the impact of simply stating quantities of pollutants in tons is likely to be several times greater than it might be in the true relationship.

202. The Primary Components of Air Pollution

The earth's atmosphere has only a limited capacity to assimilate wastes or to carry them away. Most particulate matter which is released to the atmosphere probably settles out fairly quickly; it may be that an equilibrium dust concentration in our atmosphere has already been achieved. But carbon monoxide, oxides and lead compounds do not settle out; they remain (or are transformed to other compounds by sunlight), to permanently alter the composition of the atmosphere.

Although a large number of substances pollute the air, the health hazard actually arises from the direct or indirect effect of from five to ten substances. The principal candidates for inclusion in this group are discussed below—some experts might add one or two others, some might delete one or two. Most sources of these substances can be identified, and their quantities metered on a sample basis, in ways that are not extremely expensive. For most pollutants there are a large number of control methods offering a combination of cost and effectiveness possibilities. For example, sulfur dioxide is created in the burning of fossil fuels. It can be removed from smokestack gases after burning, or removed from the fuel in the refining process, before burning. Fuels with low sulfur content can be substituted for fuels with high sulfur content. The time pattern of burning can be changed so that little high-sulfur fuel is burned when pollutant concentrations are high. Or, the location or degree of the activity for which the fuel is consumed can be altered. The health significance of a number of common air pollutants is indicated as follows:

Carbon Monoxide. The toxic effects of carbon monoxide on the body are well known. By combining with hemoglobin more rapidly than does oxygen, inhaled carbon monoxide interferes with the capacity of blood to transport oxygen to the tissues. A concentration of 30 parts per million (ppm)[3] of carbon monoxide for more than four hours will produce measurable impairment of physiologic functions such as vision and psychomotor performance. These effects are enhanced by any additional illness which decreases oxygen intake in the lungs, or the ability of the circulatory system to distribute oxygen to the body. Cigarette smokers, for example, may have carboxyhemoglobin levels as high as 8 percent. Concentrations higher than 30 ppm carbon monoxide are frequently observed in urban traffic. If this exposure is maintained for 6 hours, or for less time if exercise is carried out, significant effect on cognitive performance is also noted. Researchers Alfred C. Hexter and John R. Goldsmith have reported a significant association between death rates and the levels of carbon monoxide existing in urban communities. They found that, all other things being equal, there were 11 more deaths a day in Los Angeles when the carbon monoxide concentration averaged 20.2 ppm (the highest concentration observed during their four-year study period) as when the concentration averaged 7.3 ppm (the lowest concentration observed).

About 98 percent of the carbon monoxide that is released into the atmosphere each year comes from automobiles, which means that people may be exposed to high concentrations of the gas when they walk the sidewalks or pursue outdoor activities. Of the United States' largest cities, the highest eight-hour average concentrations of carbon monoxide have been found in downtown Chicago, with readings of 40 ppm. In New York, the eight-hour averages in midtown traffic have been in the range of 20 to 30 ppm. New standards announced by the federal government will require those figures to be reduced to no more than 9 ppm.

Hydrocarbons. Some hydrocarbons interact with oxides of nitrogen in sunlight, producing smog, ozone, eye irritation, and damage to vegetation. Hydrocarbons may have a carcinogenic (cancer-producing) effect in some compounds, but this has not been conclusively proven. The primary concern with hydrocarbon emissions is their participation in the photochemical reaction producing smog.

Nitrogen Oxides. Nitrogen dioxide is directly toxic to man and animals; other oxides contribute to photochemical reactions. The gas is relatively insoluble and can be directly inhaled, producing damage at the alveolar and lower bronchial level. Chronic respiratory disease and death have resulted from accidental exposure to high concentrations of nitrogen dioxide in mines and in farm silos. Animal studies have shown pulmonary damage with several species exposed to concentrations at 5 ppm or less; with continuous exposure, some effects have been found at 0.5 ppm. Continuous exposure also appears to increase susceptibility to bacterial infection in animals.

Most nitrogen oxides emitted from automobile exhausts are in the form of nitric oxide. A chain reaction involving hydrocarbons and activated oxygen converts the nitric oxide to nitrogen dioxide, which dissociates in sunshine to nitric oxide and atomic oxygen. Some of this atomic oxygen combines with molecular oxygen to form ozone. Ozone and related oxidents are irritating to mucous membrane, and produce eye and respiratory irritation in concentrations prevalent in Los Angeles, and occasionally in other urban areas. Evidence indicates that in the range of 0.1 to 0.15 ppm oxident (which occurs in Los Angeles), there is impairment of athletic performance of school children and more rapid impairment of lung function in persons with moderately advanced emphysema.

Lead. Lead has been known to be toxic for more than two thousand years. The symptoms of chronic lead poisoning—

weakness, apathy, lowered fertility, and miscarriage—have been recognized for several centuries. Overexposure to lead was probably a factor in the decline of the Roman Empire, because lead pipe was often used in Roman cities to carry water. Romans also lined their bronze cooking, eating, and wine storage vessels with lead to avoid the unpleasant taste of copper, thus trading the taste and symptoms of copper poisoning for the pleasant taste and more subtle poisoning of lead. Examination of the bones of upper-class Romans of the classical period shows high concentrations of lead; the lower classes lived more simply, drank less wine from lead-lined containers, and picked up far less lead poisoning.

Today in most industries and occupations which have a known hazard, lead exposure has been controlled. Most accidental exposures are by ingestion, usually through children eating lead-based paint, and through industrial exposure to fumes and particles of lead in the atmosphere. The latter may be more dangerous; only 10 percent of the lead ingested by mouth is absorbed from the gastro-intestinal tract, while about 50 percent of the inhaled lead of submicronic particle size is absorbed from the lungs.

Atmospheric lead became a major problem with the introduction of tetraethyl lead as a gasoline additive in 1924. Each gallon of gasoline contains about 2.5 cubic centimeters of tetraethyl lead, most of which is exhausted in the form of particles which remain suspended in the atmosphere. The range of blood lead is from 0.015 to 0.045 mg/100 ml of blood, with blood lead in occupationally-exposed groups such as traffic policemen ranging up to 0.06 mg. It has not been established whether this concentration is enough below the level in the blood where toxicity occurs to be safe. Persons with higher "normal" values may suffer subtle impairments of health, and perhaps impairment of enzyme systems in the body, although there may be no overt signs of disease. While it is recognized that infants are more sensitive to lead than adults, maximum permissible blood lead ranges for infants have not been established. The importance of atmospheric lead as a contributing factor to body lead and the narrow margin between normal and toxic levels in the body suggests the need for strong controls over any increased atmospheric levels in populated areas, with steps taken to reduce current levels of lead.

Carbon Dioxide. The problems associated with the production of large quantities of carbon dioxide are little understood, but have potentially vast implications. In just over a century, man has oxidized through burning a substantial portion of the earth's fossil fuels which have accumulated through millions of years. Assuming that the use of fossil fuels will continue to grow with increased economic activity, the amount of carbon dioxide in the atmosphere could rise to 150 percent of its 1970 level before the end of this century. The significance of this is not understood; it is known, however, that atmospheric carbon dioxide is one of the substances which help to retain radiated energy in the atmosphere. If measurable changes in temperature were to occur as a consequence of the carbon dioxide buildup, the effects on the world's climate and on the level of the oceans could be dramatic.

Particulate Matter. Most liquid and solid particles are submicroscopic; such particles are of 0.1 microns or less in diameter and serve as condensation nuclei which may absorb pollutants. Particles of this size acquire monomolecular layers of organic or inorganic chemicals while in the atmosphere or during passage through the upper respiratory tract. Such particles act as carriers for other pollutants and produce more serious health effects than larger particles which are intercepted in the nose or throat. Nuclei consisting of very small lead residues react with atmospheric iodine to form lead iodide, and concentrations of such nuclei may be significant in ice crystal formation. The effects of these concentrations on weather systems, notably on cloud formation and precipitation, have not yet been established but may be considerable.

Included with particulate matter are a number of substances which are potential health hazards at very low concentrations and which require stringent controls. Beryllium, for example, is emitted from industrial sources and from rocket fuel, causing lesions in the lungs, and producing serious respiratory damage and sometimes death if inhaled. Asbestos, which has long been recognized as an occupational hazard, is increasingly present in ambient air because of its use in construction materials, brake linings, and other products. Long exposure in industry produces the lung-scarring disease, asbestosis. Mercury, which has gained attention as a contaminant in tuna and swordfish, is causing concern also as an air pollutant. Mercury is often closely bound with sulphur compounds in the earth, and is freed when coal or oil is burned in power plant boilers to create steam. A high mercury content also occurs in fumes from stacks of municipal incinerators since paper is a major part of municipal refuse, and mercury also is used in the production of paper.

203. The Geographic Extent of Air Pollution

Two geographic types of air pollution—localized and generalized—are relevant both in public policy formulation, and in the consideration of specific abatement measures. A localized problem exists in a limited area—usually immediately downwind from a specific source of pollution, although factors of topography, meteorology, and the height of the stacks emitting pollutants may produce a localized problem at a considerable distance from the source. Because localized air pollution problems affect persons and property in limited segments of a community, they may be a matter of indifference to those areas of the city which escape pollution.[4]

Generalized pollution by definition affects a large area, although not necessarily large or small relative to localized pollution. The significant characteristic of generalized pollution is the complex mixing of pollutants which occurs in the atmosphere to the point where it is often impossible to distinguish one pollutant from another, or to identify their sources. Also, interactions in the atmosphere may produce secondary pollutants such as photochemical smog which differ from any of the original agents emitted.

Prevailing wind patterns and terrain in a given area determine the boundaries of an air shed, which can be thought of as analogous to the boundaries of a watershed or river valley. In California, for example, the prevailing westerly winds and inland mountains create several well defined urban air sheds, one in the San Francisco Bay area, another over and south of Los Angeles. The concept of an air shed is useful for some analytical and control purposes, although there are some problems with its use. It is much more difficult to predict air flows than water flows, and much more difficult to apply mathematical prediction tools in arriving at alternative means of air pollution abatement than water pollution abatement.

Control is made difficult also because air sheds do not generally correspond to existing political jurisdictions, and special commissions usually have to be set up and given appropriate supra-local or state powers before air-shed wide abatement can even be considered.

The upper boundary of an air shed is a ceiling resulting from lack of vertical air mixing due usually to a zone of stable temperature. Air temperature normally decreases with altitude. When temperature increases with higher altitude we have a condition known as a temperature inversion. Temperature inversions are common in all areas of the United States, occurring between 10 percent and 50 percent of the time at elevations of 500 feet or less. The intensity and duration of an inversion depend on how rapidly the earth cools at night and warms in the morning. Inversion conditions which develop over a city during the night frequently last until 3 or 4 hours after sunrise. Thus, pollution from the morning traffic peak may be held down in a city until late morning. Similarly, ground cooling may produce the beginning of an inversion at the time of the peak of evening traffic, often holding this pollution down, sometimes until morning.

204. Biological and Medical Effects of Air Pollution

A number of uncertainties in predicting the biological and medical effects of pollutants affect our ability to set long-term standards for air sheds, or to select long-term pollution control strategies. In particular, we know little of the effects of long-term exposure to relatively low concentrations of air-borne pollutants. Many of the constituents of air pollution—carbon monoxide, nitrogen oxides, and lead compounds—are well-known hazards, and toxicity standards for industrial exposure have been established for many of them. However, knowledge of the effects of moderate to low concentrations of air pollution which are continued year after year is generally not available because the necessary longitudinal studies have not been completed. If chronic or cumulative effects from prolonged exposure to relatively low concentrations of pollutants are important for certain segments of the population, such as children or the elderly, then pollution standards may have to be extremely stringent. Assessing the health effects in a human population requires continual reevaluation to better define air quality criteria. Air quality standards for pollutants in ambient air[5] have been set by state and federal regulatory agencies. However, due to the lack of health information on the effects of pollutants there should be no complacency where such standards are being met.

ECONOMIC APPROACHES TO AIR POLLUTION ABATEMENT

205. The Economic Cost of Air Pollution

Relatively unpolluted air is no longer a free good in our society. It costs money to trap pollutants before they escape into the air, and it costs money to escape to places where the air is relatively clean. Most people probably would be willing to pay for the benefits of increased longevity, decreased morbidity, and decreased nuisance that come with relatively cleaner air.

The singular characteristic of air pollution is its pervasiveness. Unlike water pollution, the extent of air pollution is limited only by the course of prevailing winds. We know that pollution of the atmosphere affects the health of human beings, of animals, and of plants;[6] it causes deterioration in property values and increased costs in a number of production processes;[7] it raises the rate of automobile and airline accidents;[8] it may substantially reduce agricultural productivity in affected areas.[9] A slow rise in the temperature of the earth has been attributed to air pollution. It is suspected of altering human genes so that mutations may occur resulting in the transmission of different characteristics to future generations.[10] Almost certainly, the major benefit from air pollution abatement is found in a general improvement in the quality of life rather than in one of the more measurable categories. It is therefore not surprising that there are fundamental problems in measuring the economic costs of air pollution.

For example, not only are current estimates of the total cost of air pollution highly speculative, but many economists in the field are pessimistic that any more accurate estimates will be forthcoming in the near future. It is possible that adequately funded studies of the costs of air pollution would be more productive than current writers seem to believe. However, the question is academic since no such studies are underway, and no funding for them appears likely in the near future.[11]

The lack of hard figures to measure air pollution costs would not be critical if there were a linear relationship between the level of air pollution and the cost of controlling it. Economists could simply make point estimates, and extrapolate to estimate intermediate values. Although some writers have assumed that the shape of air pollution cost and abatement functions are linear, it is unlikely even on *a priori* grounds that this is intended as any more than a convenient simplification. If such relationships are nonlinear, then effective policy determination requires comprehensive studies to estimate the true costs of air pollution and the benefits of abatement over a wide range of possible levels of pollution.

Harold Wolozin has proposed an S-shaped functional relationship between the level of pollution and the cost of abatement as is shown in Figure 2-1.

This is an *a priori* reasonable approximation of the likely shape of such a function.[12]

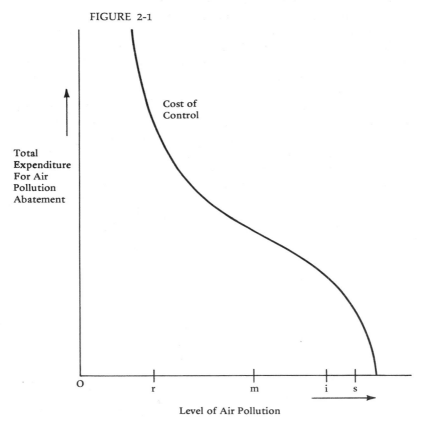

FIGURE 2-1

Cost of Control

Total Expenditure For Air Pollution Abatement

O   r   m   i   s

Level of Air Pollution

The horizontal axis on Figure 2-1 measures the various possible levels of pollution on some composite index. The vertical axis measures the total cost of abatement required to achieve each level of air quality measured on the horizontal axis. This static analysis assumes that technology is constant in the short run. The point at which air pollution is just detectable is indicated by "r;" the saturation point at which pollution is at dangerously high levels, by "s." As we move from "s" towards "r," outlays for air pollution abatement increase as we reach *successive levels* of purer air. The function indicates intitial low returns to scale as abatement is initiated at point "i," a long span over which returns to scale increase to point "m," and entry to an area of rapidly diminishing returns to scale as the air becomes cleaner.

As indicated previously, economists are concerned with an optimal level of air pollution—not perfectly pure air but some level of pollution which is acceptable in terms of the other goods that must be foregone by society to achieve the resulting cleaner air. Society accepts lower abatement expenditures—a smaller diversion of resources—at the cost of a disproportionate acceleration in the cost of air pollution.

In traditional microeconomic analysis, an optimal level of pollution would occur at that point where the total costs of pollution and of pollution control are minimized. In Figure 2-2 the Cost of Control curve is similar to that in Figure 2-1. The Cost of Pollution curve represents the cost of individual and social benefits foregone (health benefits, etc.) in the absence of air pollution control. This curve is also an S-function, with an initial range of minimal damage followed by a range of rapid rise in health and property damage costs relative to pollution levels, followed by a leveling off, although the latter might come only at a point where health costs due to morbidity and mortality were already unbearably high.[13]

The total cost of pollution and of control are at a minimum at

FIGURE 2-2   FIGURE 2-3

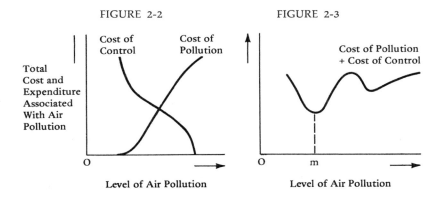

Level of Air Pollution   Level of Air Pollution

point "m" on Figure 2-3. To reduce pollution to point "m" is the same as saying that we will spend money on air pollution controls until the incremental abatement dollar reduces the cost of pollution as much as the cost of control, but that we will not spend beyond that point. The S-shape of the individual cost curves in Figure 2-2 results in a total cost curve in Figure 2-3 with several minimum points, making very difficult the estimation of a true minimum point in the absence of realistic dollar values for costs and benefits of control at different levels of pollution.

It should be emphasized that simply knowing that the total benefits from any single level of pollution abatement are greater than the total cost of pollution at that level is not a sufficient basis for deciding that abatement should be increased. It is sometimes argued in noneconomic literature that increased pollution control is necessary in a certain case because the total cost of pollution exceeds the existing industry and government expenditures for research and control devices, but this fact alone does not tell us whether pollution abatement should be increased. For example, consider the following simplified example in Table 2-1.

TABLE 2-1

Cost and Benefit of Air Pollution Abatement

Percent Level of Air Pollution Abatement	Total Benefit to Society of Abatement	Total Cost to Society of Abatement	Net Value to Society of Abatement (Total Benefit minus Total Cost)
0	$ 0	$ 0	$ 0
20	100	25	75
40	180	75	110
60	250	175	75
80	305	300	5
99	320	450	−130

The total benefits to society from 80 percent abatement of $305 exceed the total cost to society of $300 at that level. However, the net value to society of abatement is less at 80 percent than it would be at 60 percent, 40 percent, or 20 percent. If our total cost and total benefit schedules are correct, then a 40 percent level of air pollution abatement is economically optimal. The conclusion that abatement should be increased from that point because benefits would exceed costs would be economically incorrect.[14]

The two sections which follow are indicative of some of the

better economic research which has been carried out on determining the costs associated with air pollution. The first considers health costs; the second considers costs from deterioration in residential property values.

206. Health Costs of Air Pollution

Much scientific effort has been expended to investigate short-term episodes of air pollution, while the more relevant question is probably the long-term health effects of growing up in and living in a polluted atmosphere. While a number of scientists have established that air pollution is associated with respiratory diseases of many types, including lung cancer and emphysema, the qualitative link is for our purposes of limited usefulness. To estimate the benefit of pollution abatement, we must know how the incidence of a disease varies with the level of air pollution.

A 1970 article by Lester B. Lave and Eugene P. Seskin which appeared in *Science,* reported an investigation of the effect of air pollution on human health and derived quantitative estimates of the effect of air pollution on various diseases. It also discussed the economic costs of ill health, and estimated the cost of effects attributed to air pollution.[15] For example, Lave and Seskin collected data for 114 Standard Metropolitan Statistical Areas (SMSAs) in the U.S. and attempted to relate total death rates and infant mortality rates to air pollution and other factors such as socioeconomic data. These data, death rates, and air pollution data were taken from various sources, and a series of statistical regressions were run. Their data showed that a 10 percent decrease in the minimum concentration of measured particulates in the United States would decrease the total death rate by 0.5 percent, the infant death rate by 0.7 percent, the neonatal death rate by 0.6 percent, and the fetal death rate by 0.9 percent. A 10 percent decrease in the minimum concentration of sulfates would decrease the total death rate by 0.4 percent, the infant mortality rate by 0.3 percent, and the fetal death rate by 0.5 percent.

By way of illustrating the difficulty in generating these figures, consider the sources of data available to the researcher. Epidemiological data are the kind of health statistics best adapted to estimating air pollution effects. These data are in the form of mortality and morbidity rates for a particular group of people, generally segmented geographically. Thus, an analyst could in theory try to account for variations in the mortality rate among the various census tracts in a city. However, while these statistics are tabulated by the government and are easily available, there are problems both from the varying accuracy in classifying the cause of death (since not all physicians take equal care in determining causation), and from the lack of information on unmeasured variables, such as smoking habits, occupational exposure to air pollution, and genetic health factors.

A second source of data available to the researcher is from episodic relationships. These data attempt to relate daily or weekly mortality or morbidity rates to indices of air pollution during the interval in question. These studies are of limited interest because they concentrate on the immediate determinant of death rather than on the initial cause of illness. A 30-year-old who is killed by an increase in sulfur dioxide concentrations is likely to be gravely ill in the first place. Thus, morbidity data are probably more useful than mortality data.

Lave and Seskin also reviewed a number of studies which quantified the relationship between air pollution and morbidity and mortality rates, and concluded that the evidence shows a substantial association between the two. Their conclusions could be challenged on the grounds that the relationships found by investigators are spurious because the level of air pollution is correlated with a third factor, which is the "real" cause of ill health. For example, many studies do not consider smoking habits, occupational exposure, or the general pace of life. Thus, it may be argued that city dwellers smoke more, get less exercise, and tend to be more overweight, thus having higher morbidity and mortality rates than rural dwellers. Were this true, air pollution as a causative variable might be irrelevant.

Apparently there is little systemic relationship between such third factors and the level of air pollution. An English study in which smoking habits were examined revealed little evidence of differences by residence. In several United States studies, the correlations between air pollution and mortality were greater when areas within a city (where more factors are held constant) were compared than when rural and urban areas were compared. Also, significant effects were found in studies comparing individuals within strictly defined occupational groups, such as postmen and bus drivers, where incomes and working conditions were comparable and unmeasured habits were likely to be similar.

Lave and Seskin attempted to translate their data on increased sickness and death from air pollution into dollar units to answer the question of how much society should be willing to spend to improve health. The normal procedure for estimating what society is willing to pay for better health is to total the amount that is spent on medical care, plus the value of foregone earnings resulting from disability and death. However, this total would seem to underestimate the amount that society is willing to spend to prolong life or relieve pain. For example, a patient with kidney failure can be kept alive by renal dialysis at a cost of up to $25,000 per year—a cost which is substantially in excess of foregone earnings, although today many kidney patients are receiving this treatment. Lave and Seskin defined direct disease costs as including expenditures for hospital and nursing-home care, and for services of physicians, dentists, and members of other health professions, plus the earnings foregone by those who are sick, disabled, or who died prematurely.

This method of calculation can be illustrated further using the case of bronchitis. The studies cited in the Lave and Seskin article indicated that mortality from bronchitis would be reduced by about 50 percent if air pollution is lowered to levels currently prevailing in urban areas with relatively clean air. The assumption was made that there would be a 25 to 50 percent reduction in morbidity and mortality due to bronchitis if air pollution in major urban areas were abated by about 50 percent. Since the medical expenses and foregone income related to bronchitis total about $930 million per year, Lave and Seskin concluded that from $250 million to $500 million per year would be saved by a 50 percent abatement of air pollution in the major urban centers.

Similar calculations were applied to lung cancer ($33 million annual saving based on a 50 percent reduction in air pollution), all other cancers ($390 million), respiratory disease ($1222 million), and cardiovascular morbidity and mortality ($468 million). These cost estimates are not all equally certain. For example, the connection between bronchitis or lung cancer and air pollution is well documented while the connection between all cancers or all cardiovascular disease and air pollution is more tentative.

Lave and Seskin's conclusion that the total annual cost saving from a 50 percent reduction in air pollution levels in major urban areas is $2080 million or more provides a first approximation to what the health cost savings from reduction of various forms of pollution might equal.

The Lave and Seskin article and similar articles also indicate that researchers are not looking for one substance in air pollution to account for observed health effects, but rather are searching for an understanding of the complex chemistry of the atmosphere and the effects of that chemistry on human bodies. There are numerous diseases—caused in combination by infection, by hereditary predisposition, by allergy, by emotional factors, and by cigarette smoking—to which air pollution acts as a catalyst. The effect of pollutants on man is complicated by atmospheric factors of temperature and humidity, which have their own effects on health aside from their effect on pollutants in the atmosphere. Thus, in London in 1952, a number of deaths from respiratory tract ailments followed a concentration of 1.34 ppm of sulfur dioxide together with a high concentration of particulate matter and soot. In 1962 a higher concentration of 1.98 ppm of sulfur dioxide with a much lower level of particulate matter coincided with far fewer deaths. In Amsterdam in 1962, similar high levels of sulfur dioxide and much lower levels of particulate matter to that found in London coincided with virtually no increase in mortality or illness. High levels of particulate matter with low concentrations of sulfur dioxide have produced mixed results in both countries.

207. Residential Property Values and Air Pollution

There is evidence that air pollution affects residential property values, and that it may figure in people's calculations when they move. Some pollutants are undoubtedly reflected in property values; others may not be because of imperfect information. In a study by Ronald G. Ridker and John A. Henning, published in *The Review of Economics and Statistics* in 1967,[16] the authors attempted to provide evidence on the effect of air pollution on property values for single family dwellings in the St. Louis metropolitan area in 1960. The evidence consisted of estimates obtained through applying least-squares regression methods to cross-sectional data. The air pollution variable was treated by developing alternative estimates of the effects of air pollution, given other possible explanatory variables that might have been involved.

As indicators of characteristics of the property itself, Ridker and Henning included variables for median number of rooms in the house, percentage of houses recently built (as an index of housing quality), and number of houses per mile (as a measure of average lot size). As indicators of locational advantages and disadvantages, they included express bus travel time to the central business district, accessibility to highways and major thoroughfares, accessibility to shopping areas, and accessibility to major industrial areas. As indicators of neighborhood characteristics they obtained and quantified data on school quality, crime rates, persons per unit (as a measure of crowding), and occupation ratio (as a measure of the homogeneity of a neighborhood—the assumption being that people prefer to live in neighborhoods that are homogeneous with respect to occupational and social classes).

A further variable was included to indicate the effect of differences in property taxes, which are likely to be capitalized in the market value of the property. Another variable represented the percentage of non-white residents in a census tract (although no *a priori* judgments were made about what relation this might have to property values). A final variable of median family income was introduced as a proxy for the housing and neighborhood characteristics that had not been picked up by the other variables used in the study.[17]

On the whole (and considering that fifteen separate variables were involved), the hypotheses in the Ridker and Henning study tested out very well. The results were both statistically significant and fairly reasonable within the context of the St. Louis metropolitan area.

The researchers concluded that if the sulfation levels to which any single family dwelling unit is exposed were to drop by .25 mg. per 100 cubic centimeters per day (compared with a mean of 0.85, a range of approximately 0.35 to 2.75, and a standard deviation of .45), the value of that property could be expected to rise by about $245.[18] This would have produced a total increase in property values for the St. Louis SMSA of about $82 million. Invested at 10 percent, this amounts to a return of about $8 million annually. Ridker and Henning assumed it would cost about $8 million per year to shift to low-sulfur fuels that would cut sulfation levels enough to achieve the .25 mg. per 100 cubic centimeters per day level. However, other considerations are also relevant. First, property values other than those for single family dwellings would also rise, adding substantially to the benefit estimate. Second, benefits besides the increase in property values (health benefits, for example) would also be derived from a reduction in sulfation levels. Third, to bring about these property value and related benefits it would probably be necessary to reduce the levels of other pollutants that are correlated with sulfation levels, especially particulates, and this would substantially raise the cost estimates. Although the Ridker and Henning work is probably the best material to date on the relationship of residential property values and air pollution, it is clear that further work remains to be done before an adequate comparison between the benefits and costs of air pollution abatement—even when related only to changes in residential property values—can be made.

208. Control of Air Pollution

The debate among economists over air pollution control centers on whether polluters in the private sector (consumers or business) can be induced to voluntarily lessen or eliminate their pollution in response to market incentives or coercion, or whether the government must enforce control by legislative means. Under incentive systems, there is, of course, an option not to be persuaded. Wolozin quotes one businessman as follows:

> . . . if you would base pollution control on a system of incentives, you might be disappointed. The marginal dollar gained for pollution control is hardly as exciting as the marginal dollar gained in expanding sales, creating new products or improving technology . . . many if not most businesses have a shortage of key personnel and they would rather use this resource to develop the mainspring of their profits than to maximize their pollution subsidies.[19]

Market pressure could be exerted by imposing effluent fees or other charges that would reflect the marginal external costs of the air pollution. Other suggested economic incentives include tax credits for air cleaning equipment or alternative processes, outright payment by government for control devices, or government relocation cost payments.

The movement at present seems to be towards government assumption of direct responsibility rather than operation through market incentives. Except for the control of automobile emissions and the setting of overall national air quality standards, which are federal responsibilities, the establishment of enforcement regulations or incentives is a state and local responsibility. As we shall examine later in this book, both the *Clean Air Act of 1963* and its subsequent amendments, and the *Air Quality Act of 1967*, which replaced it, require states to undertake this responsibility. Prior to the Clean Air Act, only sixteen states had air pollution control legislation; today, all do.

Although the Air Quality Act gives state and local governments a primary role in protecting air quality, it gives the federal government authority to act in emergencies, responsibility to review and approve state and regional control programs, and authority to establish federal controls in states or regions which fail to establish their own controls. The Act requires the Department of Health, Education, and Welfare (HEW) to designate specific air quality control regions, which treat groups of communities as a unit for the purpose of setting and implementing air quality standards. HEW is required to develop and publish air quality criteria for pollutants or groups of pollutants, information on control techniques that identify the best methods available for reducing each pollutant emission at its source, and the cost of doing so.

After a criterion and information on control techniques for a pollutant are published, the Act sets a timetable which each state must follow in developing its own air quality standards (not lower than federal standards), and implementing them for each of the designated regions. Implementation plans can offer financial incentives as well as enforce controls. If any state fails to establish standards, or if the Secretary of HEW finds that standards are not consistent with the federal criteria, he can initiate action to insure that appropriate standards are met.

In May of 1971, the Environmental Protection Agency (see Chapter Seven) announced its first national air quality standards for six principal pollutants, to go into effect by July 1, 1975. The standards are for sulphur oxides, particulates, carbon monoxide, hydrocarbons, nitrogen oxides, and photochemical oxidants. The first two are of particular interest. The standard set for sulphur oxides is 1.03 ppm of air as an annual mean; for particulates, 75 micrograms per cubic meter. Most regions of the country can meet these standards by switching to low-sulphur fuels and by requiring plants to install electrostatic precipitators to capture soot. Seven cities—New York, Chicago, St. Louis, Baltimore, Hartford, Buffalo, and Philadelphia—are expected to have a hard time meeting the standards by 1975 because of currently high pollution rates. To meet the standards through an increased use of natural gas (replacing high sulphur coal), the seven cities combined would cause an increase in the national use of natural gas by almost 15 percent with half that increase going to New York City alone. However, as the National Academy of Engineering has pointed out, the difficulty with this solution is that the supply of natural gas will decline markedly in less than 10 years unless large new reserves are discovered.

In order that the legal deadline for carbon monoxide be met, not only would automobile manufacturers have to meet the Act's 1975 deadline of producing "clean" engines, but many cities would have to make drastic changes in their transportation systems by developing rapid transit lines from the suburbs and limiting private cars in their inner cities in peak hours.

The problem with hydrocarbons is equally severe. In many cities it is now common to find 2 to 3 parts of hydrocarbons to a million parts of air. The primary standard set by the Agency calls for a limit of 0.24 ppm as a maximum three-hour average concentration not to be exceeded more than once a year.

209. Approaches to Air Quality Standard Setting

In general, two approaches are available in the setting of air quality standards—constant pollution abatement, and selective pollution abatement.[20] Under the constant abatement solution, air quality standards are set well below the pollutant concentrations known to result in morbidity or mortality. All sources of a particular pollutant are expected to reduce their emissions in the same proportion as the desired reduction in air pollutant concentration. In one sense this is an equitable procedure, as a source that accounts for "x" percent of particulate emissions is thus responsible for "x" percent of the reduction in particulate concentration. In another sense it is inequitable, as it assumes different sources of pollution have similar costs of proportionately reducing that pollution. If the assumption is incorrect, it results in a solution which is far more costly for some polluters than others.

The best example of constant abatement is citywide or areawide standards requiring that each source of pollution reduce its emissions by a predetermined proportion. In its most naive form, this requires that a source located on the downwind side of an air shed must reduce emissions proportionately with a source located upwind thus removing any incentive that a firm has to relocate to reduce its effect on the level of air shed pollutant concentration.

Under the selective abatement solution, air quality standards are set at pollutant concentrations which, if exceeded, are known to result in morbidity or mortality. This approach recognizes that acute air pollution episodes can be predicted, and that varying degrees and methods of abatement are appropriate for different degrees of pollution. Thus, in the New York City-New Jersey Air Pollution Commission area, an initial alert is called if air pollutant concentrations exceed predetermined levels, or if stable weather conditions are forecast. A second alert is called if concentrations of air pollutants reach still higher levels, or if the first alert has not produced improved conditions. A third alert, which indicates a danger to public health, is called if the concentrations of pollutants reach even higher levels or if the second alert has been ineffective.

Successive levels of alert require that pollution control equipment be brought into use, that some polluting activities such as open burning be limited or terminated, that motor vehicle operation in affected areas be restricted, that the consumption of polluting fuels such as sulphur-heavy oil be restricted, and so on. A third-level alert might require measures as stringent as a curfew on lighting and heating (so that coal or oil fired generators can be cut back), restrictions on the use of fuel oil and diesel oil, and curtailment of all motor vehicle traffic in the metropolitan area except for emergency vehicles. This staged-approach implies among other things that companies do not have to utilize expensive pollution control equipment 100 percent of the time to satisfy air quality standards, but only when problem conditions are forecast. Further, all polluters do not have to abate to the same degree. The value of the staged-approach depends largely on how well weather conditions can be predicted. To date, meteorologists have had only limited success in forecasting air-pollution potentials accurately.[21]

Constant abatement both incurs the greatest abatement cost to society and also provides the best guarantee that there will not be an air pollution problem. Each variation in shifting from the constant abatement model to the selective abatement model is a refinement. As we move towards selective abatement there is a tradeoff; the probability of satisfying any given air-quality standard declines, while the probability of finding the least expensive solution to air pollution increases. Selective abatement carries the expected cost that air quality standards will not be met, while constant abatement carries the expected cost from not using the least-expensive approach to air pollution control. The optimal least-expected cost solution to the air pollution problem depends specifically on our knowledge of the environment and on our ability to forecast weather conditions.

Independently of the ability to forecast, general abatement will be more economic than selective abatement where the cost of implementing and supervising selective abatement is very high—where sources of pollution are small and numerous as with residential units, small incinerators, and automobiles.

The most efficient solution will also differ between cities or air sheds, each of which has its own unique characteristics and problems; each must determine for itself whether constant abatement or some variation of selective abatement is most efficient in meeting specific air quality standards. This means it is probably uneconomic for the federal government to establish national air quality standards or emissions standards, except perhaps at very minimal levels, or as guidelines.

AUTOMOTIVE AIR POLLUTION: A CASE STUDY

210. A Systemic Approach to Automotive Air Pollution

We have already noted that the automobile is the biggest single contributor by weight to air pollution. Some conclusions from a study commissioned in 1966 by the United States Department of Commerce provide further perspective on this major pollutant. The study, led by Dr. Richard Morse of MIT, was originally to consider the feasibility of electrically powered vehicles. The resulting Panel on Electrically Powered Vehicles subsequently broadened the scope of their inquiry to include other modes of transportation, and to consider the total complex of urban problems related to automotive air pollution.

The Morse study used a systemic approach to point out where new solutions might be fruitful. The systemic approach is illustrated by the observation that pollution control changes may affect more than just vehicles. In evaluating the effect of changing permissible pollutant levels, one must also examine possible impacts on transportation system design. For example, emission reduction would facilitate the design and construction of tunnels, and partially or totally underground highways. Ventilation requirements limit the extent of both tunnels and underground roads, and increase the cost of building and operating them. The reduction of ventilation requirements that would result from reductions in pollution emissions would free designers from many current constraints.

Also, emission reduction accompanied by noise reduction would permit extensive use of the air space over highways. In many cities, roads and access ramps use large amounts of land and destroy neighborhood integrity. Development of highway air space might offset these effects, but such development is likely to take place only if automobile emissions and noise are substantially reduced.

The concluding chapter of the Morse report conceives of vehicular air pollution as only one element in a complex of urban problems that begin with the existing spatial relationship of homes to jobs. There has been considerable debate as to whether this systemic approach is a feasible short-term one to the problem of vehicular air pollution, or whether the scope of the problem requires a more pragmatic short-term consideration only of the cost-benefit aspects of surface vehicle propulsion systems. For example, does the ultimate solution to automotive air pollution problems lie only with the design of vehicles powered by steam or electricity, which are attractive to the consumer and inherently free of toxic emissions? More basic, is surface transportation in urban centers a determining force, or can it be treated as a dependent variable which responds to external criteria? The conclusions of the Morse report are stated in the following excerpt.

THE AUTOMOBILE AND AIR POLLUTION*

One might reduce the level of polluting activity by
(a) Reducing the total amount of urban transportation.
(b) Reducing the total amount of motor vehicle travel throughout a metropolitan area
(c) Shifting some of the demand for motor vehicle travel to less polluting modes of transportation.
(d) Reducing total motor vehicle travel in severely affected areas

---

*Reprinted from Richard Morse, et al., The Automobile and Air Pollution; A Program for Progress (The Morse Panel), Report of the Panel on Electrically Powered Vehicles, Part II, Washington, D.C.: U.S. Department of Commerce (October, 1967). Reprinted by permission of Richard S. Morse, Chairman of the Panel. Portions of the original sections have been omitted.

(e) Reducing all motor vehicle travel in affected areas at affected times.

(f) Reducing the extent of motor vehicle travel in especially polluting aspects of driving (with possible further qualification to specific areas and/or times).

(g) Reducing travel by more highly polluting vehicles.

Reducing The Total Amount of Urban Transportation

There are basically two ways to reduce the total amount of urban transportation: to reduce the demand and/or to restrict the supply. To reduce the demand, two main approaches have been proposed. The first is to redesign sections of metropolitan areas to reduce need and demand for transportation, and to so design new areas. The second is to substitute communications services for direct transportation.

• • •

The lengths and frequencies of trips in a metropolitan area are functions of its structure as well as of its transport system and socio-economic characteristics. The relationships of homes to jobs, and to shopping, cultural, educational, and recreational facilities strongly influence the over-all amount of metropolitan travel. It has thus been suggested that the total demand for transportation might be reduced if one could

(i) Provide better spatial integration of residences and offices in communities, to reduce the home-work place separation. At one extreme, this could involve recreating villages or village-like environments that would be relatively self-contained. . . . A more moderate version of the same proposal might involve revising zoning practices to permit greater intermixing of residences and offices throughout metropolitan areas.

(ii) Substitute elevator for ground transport by judicious use of tall buildings. Tall buildings abet urban concentration, which has been both strongly supported and strongly attacked in the last few decades. . . .

(iii) Redesign Federal regulations and financial procedures to encourage developments containing homes, apartments, and jobs, to minimize the number of new "bedroom" communities miles away from places of employment.

The substitution of communications services for direct transportation has been a popular theme in projections for the future. It has been forecast that phonovision, closed-circuit television, fac-

simile transmission, remote computer operation, and devices yet
unforeseen will make travelling all but obsolete. Yet evidence thus
far suggests that transportation and communications reinforce each
other: the better the communication the more transportation, and
vice versa.

•   •   •

Reducing The Total Amount Of Metropolitan Motor Vehicle Travel

Basically, there are three ways to reduce the total
amount of metropolitan motor vehicle travel: regulation, pricing,
and shifting demand to other transport modes.

Regulation might involve (a) fuel rationing, using any of a
graded spectrum of schemes, (b) limited registration of second and
third cars (based on "evidence of need") in the metropolitan area,
(c) extremely tight enforcement of traffic laws, (d) motor vehicle
travel permits (i.e., rationed travel), etc.

Pricing might involve (a) substantially increasing fuel taxes
throughout the metropolitan area and its environs and (b) substantially increasing vehicle registration and operator's license fees
throughout the area, to reflect the cost of air pollution in the
expense of automobile operation. In conjunction with (c) above, it
might also include increasing fines for traffic violations.

The pricing schemes tamper less with normal economic and
social functions but may lead to misallocation and maldistribution
of resources without successfully reducing pollution. The schemes,
for example, are all economically regressive. They thus affect most
those with lower incomes, who own fewer vehicles and drive those
they do own less than the overall average.

•   •   •

Shifting Demand To Less-Polluting Modes Of Transportation

Electric Vehicles. . . .Electric cars need be considered
for pollution-reduction only over the long-term. They are not
expected to have any significant impact until at least 1980.

. . . If the new cars are truly as different and new as their most enthusiastic advocates claim, they will radically change the relationship between existing transportation modes and tend to create their own specialized markets.

Mass Transportation. Mass transportation must be effective at the margin if it is to be effective at all in reducing pollution. It might be effective, for example, if it were to discourage additional one-car families from becoming two-or-more-car families. It might be effective if it were to remain sufficiently attractive and convenient to keep the passengers it now has. And it might be effective if its diversion of demand were to be sufficient to reduce peak-hour congestion in some areas and thus reduce the pollution caused by vehicles operating in highly inefficient driving modes.

Mass transportation as now conceived is not likely to be able to do much more. No matter how successful mass transportation is, it will not handle goods, which will continue to be carried mostly by truck. And, perhaps more important, mass transportation, even where it is expanding, is not the only form of transportation being built. New highways and highway improvements are underway or planned in many areas, and no cessation of road construction or improvement appears likely.

In many urban areas, mass transportation now carries an important fraction of people commuting to downtown. The question of interest for air pollution, however, is how many of those not now using mass transportation can be induced or persuaded to shift. For most urban areas, this number appears to be small, relative to the total number of people driving motor vehicles in the area for one reason or another.

With regard to the new San Francisco Bay Area Rapid Transit system, for example, the AAAS Commission on Air Conservation found: "Roughly 100,000 automobile trips per day are to be diverted to rapid transit. From the viewpoint of coping with traffic and congestion at the hours of most intensive demand, rapid transportation could make an important contribution. From the viewpoint of air pollution control, however, the transit solution, even if it meets the expectations of the Transit District, is only a minor portion of the whole. There are now [1965] about 4 million daily trips in the three-county area, 7 million in the entire Bay Region. These figures [are expected to] rise to 5.2 million and to 11 million, respectively, by 1975. The Transit District estimate that it would carry 258,600 trips per day means that it would absorb only about 5 percent of the passenger travel (in trips, not miles) in the three-county area, and 2.5 percent in the entire Bay Region."

Reducing Motor Vehicle Travel In Affected Areas

There are three basic ways of reducing motor vehicle travel in affected areas: regulation, pricing, and shifting demand to other modes.

Regulations that have been proposed include:

(a) Banning or regulating the use of internal combustion engine vehicles in affected areas (possibly only at certain times).

(b) Restricting traffic access to major arteries, bridges, tunnels, etc. leading into affected areas (possibly only at certain times).

(c) Reducing the number of parking spaces available for all-day or long-term parking in affected areas.

Pricing policies that have been proposed include:

(d) Instituting graduated, variable tolls on major arteries, bridges, tunnels, etc. leading into affected areas. Such tolls could be adjusted to help control traffic patterns and reduce traffic into affected areas at selected times.

(e) Substantially increasing long-term parking charges in areas used by commuters, while keeping charges relatively low for short-term (less than 4 hours) parking to encourage shopping, social, cultural, and recreational activities.

(f) Indirectly achieving (e) through selective taxation and legislation.

(g) Instituting vehicle use taxes (similar to wage taxes), to be collected from those who use vehicles regularly in affected areas.

Policies for shifting demand to other modes include:

(h) Improving public transportation in affected areas.

(i) Making non-polluting vehicles available for inexpensive rental, in conjunction with one or more of (a) through (g).

(j) Making parking spaces more available for all-day parking near collection points for non-polluting modes of transportation ("Park-and-ride").

(k) Improving taxi availability and service, in conjunction with (h) and (j).

The regulations restrict mobility and may reduce pollution at the expense of reducing the activities that attract people into affected areas. Good analysis and study would be needed to assess the regulations' impact.

The pricing policies have attractive aspects, but have the major disadvantages of being somewhat regressive and of working only indirectly to reduce pollution.

Reducing The Extent of Driving In Inefficient Modes

There are basically two ways of reducing the amount of driving in inefficient, highly polluting modes: introducing new vehicle configurations and concepts and improving traffic flow.

The former reduces inefficient driving by substituting non- or low-polluting power sources for the internal combustion engine. The main new vehicle configurations and concepts proposed are:

(a) Combination of car or truck powered by internal combustion engine with auxiliary transport system to reduce mileage driven— e.g., piggyback, moving pallet, road-rail system.

(b) Combination of short-range electric vehicle with long-range transport system—e.g., piggyback, "third-rail," electronic highways, "Urbmobile."

Improving traffic flow helps reduce pollution because vehicular pollution is created at a rate approximately inversely proportional to speed. An increase in speed from 20 to 30 miles per hour yields about a one-third reduction in pollutants, while a change from 20 to 40 miles per hour brings about a two-fold reduction. There is considerable room for improvement: Traffic on Manhattan streets during rush hour moves at an average speed of 8½ miles per hour, and on the approach expressways to Manhattan, speeds are as low as 13 miles per hour.

The two basic approaches to improving traffic flow are

(a) reducing traffic levels. Some ways are staggering working hours and encouraging a greater number of persons per individual vehicle.

(b) improving traffic flow at given levels of traffic. Some ways include better traffic control, new expressways, redesigned intersections and traffic bottlenecks, automatic control devices to govern individual vehicles (automated highway), better system labelling.

Reducing Travel By More Highly Polluting Vehicles

Some vehicles emit more pollutants than others. One might thus reduce average emissions per vehicle by reducing the proportion of travel done by more highly polluting vehicles.

Policies proposed for accomplishing this include:

(a) Imposing taxes or "polluting fees" related to emissions. . . .

(b) Causing polluting vehicles to be operated more efficiently. This might involve checking all cars and trucks for combustion efficiency, training mechanics better to service and maintain pollution-related systems. . . .

(c) Having motor vehicle taxes increase with increasing vehicle age.

Although some old cars burn fuel more completely than some new ones, statistically age is highly correlated with poor carburetion. Approximate, preliminary analysis indicates that this tax scheme would affect mainly the disadvantaged and poor, who tend to own older cars, and middle-income families owning two or more cars. In the first case, the taxes would be regressive and would seem unlikely to achieve much of the desired effect. In the second case, the increased tax might stimulate earlier "trading in" of old second cars but would drive off the road primarily cars being used far less than average.

• • •

Alternatives to Reduce The Impact Of Emitted Pollutants

There are basically two approaches to reducing the impact of emitted pollutants: reducing atmospheric pollutant concentrations in areas where they may affect people, property, or plants and reducing the impact of the pollutants on the sensitive receptors.

Policies proposed to reduce atmospheric concentrations include:

(a) Reducing local concentrations by :

(i) Enhancing local atmospheric transport, to disperse pollutants —e.g., enhancing natural micro-meteorological forces by shaping wind patterns (building location and design, major topographical changes, such as building or levelling hills), making best use of the winds available (highway location and design), and reducing the extent of construction (and forest denudation, etc.) favorable to stagnant conditions .

(ii) Providing artificial convection.

(iii) Preventing pollutants from accumulating in local atmospheres, by "sweeping" with adsorbents, molecular sieves, or reactive catalysts to remove pollutants as they are formed.

The forces involved in (i) are not well understood, but the possibilities appear attractive. Achieving (ii) is likely to be inordinately expensive; the atmosphere ordinarily tends to resist or overcome small-scale disturbances, and even assuming atmospheric cooperation the power and equipment costs involved are enormous. Alternative (iii) involves treating large volumes of air, but may be very attractive and effective in places where air is already collected and pumped as, for example, at tunnel ventilators or garage exhaust vents.

(b) Removing undesirable products from the general atmosphere:

(i) Chemically—react pollutants to form carbon dioxide and water or other less desirable compounds.

(ii) Physically—sweep, scrub, or scavenge pollutants from atmosphere. Enhance precipitation of particulates.

(iii) Biologically—introduce pollution predators, such as microbes or insects.

(i) and (ii) require contacting large volumes of air with other substances. The power needed simply to move the air effectively is likely to be so expensive that these alternatives will be most unattractive except where the air is moved anyway, as in tunnel ventilation.

The problem with (iii) is uncertain ecology. It is difficult to predict what else pollution predators might find appetizing, or what effects they might have on their surrounding biological system. (The "cure" may be worse than the problem.) It also may be difficult to "train" predators to feed only on undesirable substances.

(c) Reducing megalopolitan sprawl.

Through "downwind" and cumulative effects, filling in metropolitan areas is believed to exacerbate pollution. Areas that once received relatively clean air receive pollutants from new sources in filled-in areas upwind; these new pollutants compound the older area's existing problems. And pollutants that once were blown quickly out of the megalopolitan convection cells are believed to stay much longer when the megalopolis expands.

Policies proposed to reduce pollutant impacts include:

(a) Locating major highways such that pollution effects on surrounding neighborhoods are kept to a minimum.

(b) Reducing the biological effects of atmospheric pollutants on man by:

(i) Relocating pollutant sources relative to man so that the effective exposure is less.

(ii) Designing and distributing protective systems—e.g., gas masks, oxygen, anti-pollutant inoculations.

(iii) Making effective diagnosis and treatment of pollution effects readily available.

(iv) Developing a means for inducing tolerances and immunities to pollution effects.

(i) is likely to become quite important, although the design bases for it are ill-understood. (ii) may become necessary for especially susceptible people (i.e., heart patients, elderly people with respiratory ailments) in high-pollution areas. The need for (iii) and (iv), and the scientific bases on which to develop them, are still unclear.

Impacts Of Pollution Control Policies

Recommendations . . . that would result in significant reduction of air pollution are likely to change the pricing cost, performance characteristics, and accessibility of the transportation system. These changes could result in significant net costs or benefits to society.

Pollution control affects not only the transportation system but also products—automobiles, trucks and motor fuel—that account changes resulting from pollution control measures, therefore, could for a large part of the American gross national product Major have large impacts on the national economy.

If, to choose an extreme example, electric vehicles were to predominate some decades hence, vast amounts of petroleum now used to make gasoline would have to be converted into other products or be left unrefined or underground. Service stations would have to be converted to handle the new vehicles and their particular requirements. Electrical distribution networks would have to be extended and expanded in heavily populated areas to permit the distribution and use of recharging outlets. Batteries or fuel cells (or other energy conversion devices) would have to be designed to minimize the use and/or facilitate the recovery of scarce or precious materials.

211. Market Structure and Low-Pollutant Propulsion Systems

The possibility of electric or steam-powered automobiles representing a viable alternative to the internal combustion engine must be considered against the background of the market structure of the automobile industry. For several decades, three to five companies have dominated the domestic production of automobiles. The barriers to entry in the automobile industry are formidable. For example, economies of scale are such that to produce at near minimum cost, a company must turn out 250,000 or more units of the same model per year. The capital investment required for production of this magnitude is from $75 to $125 million. Also, a strong and expensive dealer system is required to market and service automobiles. Investment per dealership runs upward of $125,000 with $350,000 a minimum figure for a large dealership. Perhaps 200 such dealerships are required for minimum national coverage; the automobile manufacturer has the options of making this investment directly, or inducing dealers with franchises to make the investment against possible returns from sale of an untried product. Exclusive dealing arrangements prohibit most existing automobile dealerships from taking on competing product lines without permission from their parent franchisor. In addition, yearly model changes have been an important part of marketing strategy for American automobile manufacturers. New and smaller firms are at a distinct cost disadvantage in this area since the substantial retooling costs involved in such changes must be spread over the number of units produced during the model run.

Any substantial innovation in the automobile industry would require a serious reorientation in industry production, materials usage, and relationships with subcontractors and supporting industries. A steam-powered vehicle would use the same production and service facilities as do internal combustion vehicles, but would make obsolete existing transmission, ignition, and cooling systems and would require a completely new oil refining process to meet the fuel requirements of steam engines. However, an electric-powered vehicle would also make obsolete engine, power train, cooling system, and existing gasoline fuel supply facilities. It is doubtful that the automobile industry, as presently structured, would pursue either of these innovations in the absence of a strong competitive threat.[22]

Steam propulsion systems do not appear to have the potential for reducing the barriers to entry into the automobile industry. In all relevant aspects of production and marketing, a steam propulsion system would be close enough to existing internal combustion systems to utilize current facilities. A steam-propulsion innovator external to the existing automobile industry would require a fairly high rate of production for minimum cost. The possibilities of subcontracting for components is limited. If the innovator overcame the capital and production barriers and marketed an attractive product, he would likely face short-run competition from the existing industry which could adapt its production and service facilities to steampower if forced to by competition. It is doubtful that the capital investment required would be forthcoming when the potential payoff is so doubtful.[23]

On the other hand, the electric automobile does seem to have the potential for reducing barriers to entry into the automobile industry. Small economies of scale, and the ability to subcontract for major components such as motors, batteries, and control systems would greatly reduce the required amount of capital to produce electric cars. Low maintenance requirements and the estimated 100 mile range of electric vehicles would permit a simpler dealer network which could initially be limited to a few major urban centers.

The likelihood of a new product image for the electric automobile suggests that the importance of dynamic obsolescence as a marketing concept will be greatly reduced, at least initially. Since the barriers to entry are sharply reduced, a small innovator has an opportunity to survive all but the most predatory behavior on the part of existing manufacturers. Thus, while steam-propelled vehicles probably constitute less of a threat to existing market structure than do electric vehicles, entry barrier considerations suggest that an electric automobile innovation is the more likely of the two in the short-run.[24]

A third low-pollutant propulsion system, the gas turbine, should also be mentioned. The gas turbine has been a favorite experimental toy of various manufacturers, notably the Chrysler Corporation. While a gas turbine does not burn externally in the sense that a steam engine does, it does burn continuously, and therefore is cleaner than an internal combustion engine. Until recently, it was felt that heat and nitrous oxide emission problems with small turbines would make them ultimately impractical for passenger car adaptation. General Motors and Ford have concentrated on larger engines for use in buses and trucks, some of which are currently operating and are close to being competitive with diesel engines on a dollar-per-horsepower basis. The passenger car turbine is a logical next step. While the turbine shares or exceeds the high barriers to entry of the steam car in the automobile industry, it is the most likely of the low-pollutant propulsion systems to be introduced by existing automobile manufacturers. However, a great deal of work to reduce costs and operating problems will be necessary.

There is some chance that the internal combustion engine will become virtually pollution-free. Edward N. Cole, President of General Motors Corporation, has predicted that by 1974 GM would have available devices capable of reducing present air pollutants such as unburned hydrocarbons, carbon monoxide, and oxides of nitrogen to near zero. He said that this would be done with catalytic converters to convert nearly all hydrocarbons and carbon monoxide to water and carbon dioxide, and cutting sharply the oxides of nitrogen emissions. He emphasized that this would only be possible with unleaded and moderately high octane gasoline which the petroleum industry is beginning to develop.

212. Federal Standard Setting for Automotive Emissions

The federal emission standards which we have noted were initially set for motor vehicles in terms of ppm concentrations of the various pollutants found in the exhaust of vehicles. Such a measurement does not take into account the total volume of pollutants and might equate small engines with larger ones which consume far more fuel and expel more total pollution. The most recent federal standards are stated in terms of grams per mile, which accounts for differences in vehicle size and fuel consumption.

Testing is done on prototype cars provided by manufacturers, and the cars are averaged so that some may exceed the standards but be offset by others which are within them. This procedure has been criticized as has the use of production prototypes; critics maintain that the cars provided for testing are highly-tuned and do not approximate production-line models in their emissions characteristics. Also, once the new car has been driven for a short time its emissions tend to increase as the car goes "out of tune."

The federal government concerned itself first with crankcase hydrocarbon emissions—the PCV valve reduced these by 100 percent.

The next stage was to reduce the permissible emission of hydrocarbons and carbon monoxide from the exhaust. 1971 models were
the first to control evaporative emissions of hydrocarbons from
the carburetor and gas tank. In 1973, controls will be extended to
nitrous oxides; in 1975, to particulate emissions.

The new standard for particulates is designed expressly to exclude leaded gasoline, which is the principal source of particulates
in the exhaust. It is considered impossible to reduce particulate content of exhaust sufficiently to meet the 1975 standard, using leaded
fuels. However, some petroleum companies have produced reduced-
lead content fuels which might make the 1975 standard. The publicly stated reason for the exclusion of lead is to permit use of
catalytic devices which may be effective in reducing gaseous emissions, but would be fouled by leaded fuels. The standard for particulate emission was set at 0.1 gram per mile, and the automobile
industry was given five years to develop the hardware to meet this
theoretical standard.

Table 2-2 shows the evolution in emissions standards through
1975. It is predicted that standards for 1980 will approximately
halve the 1975 permitted emissions.

TABLE 2-2

| | (grams per vehicle mile) | | | | | |
	Uncontrolled	1968	1970	1971	1973	1975
Exhaust:						
Hydrocarbons	12.2	2.4	2.2	2.2	2.2	.5
Carbon Monoxide	79.0	35.1	23.0	23.0	23.0	11.0
Nitrous Oxides	6.0	6.0	6.0	6.0	3.0	.9
Particulates	.3	.3	.3	.3	.3	.1
Crankcase Hydrocarbons	3.7	0	0	0	0	0
Evaporative Hydrocarbons	2.8	2.8	2.8	.5	.5	.5
Totals:						
Hydrocarbons	18.7	6.2	5.0	2.7	2.7	1.0
Carbon Monoxide	79.0	35.1	23.0	23.0	23.0	11.0
Nitrous Oxides	6.0	6.0	6.0	6.0	3.0	.9
Particulates	.3	.3	.3	.3	.3	.1

The 1971 standards reverse the steady upward trend in total
emissions, but only for a short period as the number of vehicles
and their use are increasing rapidly. By 1980 increased total mileage

will raise total emissions above present levels, assuming that the 1980 standards are half those of 1975.

The federal emission standards are also a point of friction between the United States and Europe and Japan, with Europeans and Japanese fearing that the legislation will be used as a nontariff barrier to reduce their share in the American automobile market. The stakes are high; Volkswagen sells half of its output in the United States, while companies like Volvo, British Leyland, Fiat, and Renault have smaller but significant shares of the export market. In part the problem arises because the 1975 standards are thought by the Europeans and Japanese to be way below anything that could possibly be injurious to health—with an official of British Leyland going so far as characterizing the American anti-pollution drive as "hysteria." [25]

European manufacturers have done little work in gas turbines, with the bulk of their research efforts going into electric or battery-operated cars. This technology is better suited for Europe, where distances covered by a car are much shorter than in the United States. With Europeans concentrating on electricity, and the possibility existing that U.S. manufacturers will proceed with gas turbines or some variation of the steam engine, there is a chance that European and American cars might have completely different types of propulsion systems in the future.

REFERENCES

[1] A simple counting of pollutees is misleading; there is some evidence that a relative absence of air pollution is a superior economic good—one which varies directly with real income. A study by the U.S. Public Health Service in Clarkston, Washington, a city with a pollution-producing kraft pulp mill, found that 44 percent of the local managers, proprietors, and professional people (who one would expect to have higher incomes and wealth) were "aware" of an air pollution problem and "concerned" about its effect on their health and property values. Only 32 percent of the clerical and skilled laborers, and 19 percent of the semi-skilled and unskilled workers expressed a similar awareness and concern. U.S. Public Health Service, *Community Perception of Air Quality: An Opinion Survey in Clarkston, Washington* (Washington, D.C.: U.S. Public Health Service, Publication 999-AP 10, 1965), pp. 45-55.

[2] Senate Committee on Public Works, *A Study of Pollution—Air*, 88th Congress, 1st Session (1963), vii. at 2.

[3]In talking about pollutant concentrations of parts per million (ppm), the reader should appreciate that serious health effects arise from extremely small quantities of pollutant. One part per million is roughly equivalent to an ounce of vermouth in 7,550 gallons of gin.

[4]Treatment of localized problems is discussed in Lester Goldner, "Air Pollution Control in the Metropolitan Boston Area: A Case Study in Public-Policy Formation," in Harold Wolozin, ed., The Economics of Air Pollution (New York: W. W. Norton & Company, Inc., 1966), pp. 127-161.

[5]The assumption of ambient air is itself part of an uncertainty. Knowledge of small-scale atmospheric convection and diffusion, and of the governing processes, is essential in relating air purity standards to the biological and physical effects of pollutants. Where pollutants are rapidly dispersed, higher emissions are acceptable. Where air is stagnant or atmospheric inversions are common so that pollutants accumulate, the most stringent standards are necessary.

[6]In general see J. R. Goldsmith, "Air Pollution," in A. Stern, ed., Air Pollution and Its Effects (New York: Academic Press, 1968), p. 547.

[7]There is great uncertainty on the quantitative effects of pollution on animals, plants, and materials, with many of the estimates only quasi-informed guesses. Thus, the most quoted figure for annual air pollution damage in the U.S. is $11 billion, which is derived from an estimate of costs in Pittsburgh in 1913. Economists calculated that smoke damage in Pittsburgh in that year, for cleaning, maintaining and lighting homes, businesses, and public buildings, amounted to $20 per capita per year. This figure was adjusted to 1959 prices and the updated estimate multiplied by the 1958 U.S. population to arrive at the figure of $11 billion. The President's Council on Environmental Quality estimated in 1971 that air pollution costs the United States $16 billion a year, made up of $6 billion in human mortality and morbidity, $4.9 billion in damage to crops, plants, trees and material, and $5.1 billion in lowered property values.

[8]Public Health (Johannesburg) No. 63 (1963), pp. 30.

[9]H. Wolozin and E. Landau, "Crop Damage from Sulphur Dioxide," Journal of Farm Economics,No. 48 (1966), pp. 394.

[10]Remarks of Dr. Haagen-Smit before the American Meteorological Society, Washington, D.C., January, 1968.

[11]Ronald G. Ridker completed in 1966 a detailed study of the economic costs of air pollution under the sponsorship of the Division of Air Pollution of the U.S. Public Health Service. Ridker considered the economic costs of diseases associated with air pollution; the cost of soiling

and materials damage; the cost of declining property values in polluted areas; and psychic costs and attitudes associated with a particular pollution episode in Syracuse, New York. His work is worthy of attention both as a first estimate of the value of benefits to be gained from expenditures on abatement, and for his discussion of strategies for measuring the costs of pollution. Reference: Ronald G. Ridker, *Economic Costs of Air Pollution: Studies in Measurement* (New York: Praeger Publishers, Inc., 1967).

[12]Harold Wolozin, "The Economics of Air Pollution: Central Problems," *Law and Contemporary Problems,* No. 33, pp. 229-233. An interesting illustration of curvilinear total damage and total cost of control functions with figures to illustrate their derivation is given in Azriel Teller, "Air Pollution Abatement: Economic Rationality and Reality," *Daedalus* (Fall, 1967), pp. 1085-1088.

[13]Cost and abatement curves may both be kinked, which means that the derived marginal curves would be discontinuous, introducing more uncertainty into the calculation of optimal levels of expenditure on air pollution. Each of these possibilities justifies increased research on determining the actual position and shape of cost and damage functions. Collection of data would be done separately for each specific air shed, and for specific pollutants (or groups of pollutants if they can be acted on together). Each community has its own characteristics and faces different meteorological, topological, and economic conditions. The appropriate tradeoffs will thus vary with each geographic area and pollutant being studied, which suggests that national estimates of pollution and control costs would not be useful for standard setting except as a first approximation, even if they could be made in marginal terms.

[14]The argument is developed in Azriel Teller, "Air Pollution Abatement: Economic Rationality and Reality, "*op cit.,* pp. 1087-1088.

[15]Lester B. Lave and Eugene P. Seskin, "Air Pollution and Human Health," *Science,* Vol. 169 (August 21, 1970), pp. 723-733.

[16]"The Determinants of Residential Property Values With Special Reference to Air Pollution," *The Review of Economics and Statistics,* Vol. 49 (1967), pp. 246-257.

[17]The statistical problems faced were also important. In brief, if a model has been properly specified, then least square estimates will be unbiased. If, however, a variable that *a priori* judgment suggests should be included in the analysis is omitted, the regression coefficients for the remaining variables with which it is correlated will be biased. Biased estimates will also occur if a variable that *a priori* judgment suggests should be excluded is for some reason included in the regression analysis. The extent of such biases due to incorrect specification of the

model depends upon the degree of correlation between the variable incorrectly excluded or included, and the variables whose coefficients are critical to the analysis. There are no set methods for detecting and treating this problem; the Ridker and Henning analysis involved stepwise regressions to observe the effect on regression coefficients when new variables were included, and correlations of each independent variable against all others to observe the magnitudes of their multiple and partial correlation coefficients.

[18]An earlier study carried out by Professor Crocker and Anderson in St. Louis, Washington, and Kansas City, found that a 5 to 15 percent decrease in air quality correlated with a $300 to $700 decrease in property values.

[19]Quoted in Wolozin, "The Economics of Air Pollution: Central Problems," op. cit., p. 236.

[20]The categories can be broken down further. See Azriel Teller, Air Pollution Abatement: An Economic Study Into The Cost of Control (Ann Arbor, Michigan: University Microfilms, 1968); and Teller, "Air Pollution Abatement: Economic Rationality and Reality," op. cit.

[21]Teller reports that using one set of criteria over a one-year period, twelve stagnation cases occurred of which ten were forecast. Another eight were forecast but not verified. Source: Teller, "Air Pollution Abatement," op. cit., pp. 1092-1095.

[22]Ford Motor Company is engaged in a joint venture with Thermo Electron Corporation of Waltham, Massachusetts in designing an organic working fluid Rankine-cycle steam engine for automotive application. The target date for a go, no-go decision is 1975. See Report by Thermo Electron to Division of Motor Vehicle Research and Development of the National Air Pollution Control Administration (June, 1970).

[23]William P. Lear, who has spent about $12.5 million on steam-powered experiments, admits the steam engine is five times as complicated, burns 50 percent more fuel, is twice as heavy and twice as expensive as a gasoline engine of similar power output. The automotive steam engine has five different systems: boiler, feedwater pump, expander, condenser, and controls. There are still technological problems in each system; the prime problem exists with the lack of an efficient and compact condenser.

The steam engine does promise cleaner air and fewer automotive emissions. According to General Motors Corporation and based on results from its two experimental steam cars in May, 1969, the steam car emits 0.62 grams per mile of hydrocarbons while the standard 1970 V-8 powered car emits 2.2 grams per mile. Carbon monoxide emissions from a steam engine are 2.8 grams per mile against 23 grams for the gasoline-powered V-8. Oxides of nitrogen emitted from the steam engine are about 1 gram per mile, versus about 4 grams per mile for the gasoline engine. While distinctly lower, these steam car emissions would not meet proposed 1975 emission standards for oxides of nitrogen, and would not meet proposed 1980 standards for emissions of hydrocarbons.

[24]Robert Ayres of Resources for the Future, notes that if electricity were generated totally by fossil fuels, emission of oxides of nitrogen from an all-electric car stock would exceed those from a steam-powered fleet, but would be only one-quarter those from an internal combustion engine fleet. Sulphur dioxide would present a greater emission problem also. If nuclear fuels were used, the problems of radioactivity and thermal pollution would be aggravated but not proportionately because of the use of off-peak power for recharging.

[25]An exception is Volvo, which has agreed to adopt the American standards timetable to its own cars in Sweden.

3

Water

Pollution

Rub a dub-dub
Three men in a tub
And who do you think they be?
The skipper, the shipper
The nautical dripper,
Who spill oil on Flipper,
And foul up the sea.

300. The Nature of Water Pollution

Water pollution is generally more localized than air
pollution, and sources of water pollution are generally easier to
identify and control than sources of air pollution. The problem is
still critical, however, since humans require water as well as air for
survival.

A complete description of the physical, chemical, and biological
aspects of water pollution would require an entire volume. One
common classification system of water pollution differentiates between physical wastes, which may be inorganic, organic, radioactive, etc., and biological substances, such as bacteria and viruses.
Another classification which is more useful for our purposes, dis-
tinguishes between nondegradable and degradable wastes based on
their behavior in receiving waters. Nondegradable wastes are those
which are diluted but are not appreciably reduced in weight in the
receiving waters. Degradable wastes are reduced in weight by the
biological, physical and chemical processes which occur in the receiving waters.[1]

Nondegradable Wastes. The combination of indus-
trial waste, agricultural irrigation, and mine discharges presents the

major nonnatural source of chlorides and metallic salts of local and regional waters. Nondegradable wastes are composed mainly of inorganic chemicals such as chlorides, synthetic organic chemicals, and inorganic suspended solids. A principal source of these wastes is industrial discharges which frequently contain inorganic or metallic salts, synthetic organic chemicals, and similar materials. Mining discharges may contain residues of copper, zinc, uranium, and other compounds. Acid drainage from mines is an acute problem in a number of coal producing areas. Agricultural irrigation also contributes nondegradable wastes, as the return flow from agricultural irrigation is generally much higher in dissolved salts (and chemicals from fertilizers) than was the original irrigation water.

There are several natural sources of chlorides and dissolved solids. Natural formations such as salt deposits result in high chloride concentrations in nearby groundwater and rivers. An increasingly important source of natural chlorides is seawater intrusion into groundwater near coastal areas following excess pumping from wells which results in a lowering of groundwater levels.

Nondegradable suspended solids consist primarily of sediment from natural and accelerated erosion of land surfaces and stream channels, and of colloidal clay particles from domestic and industrial wastes. In particular, storm runoff from agricultural land carries silt, clay, and fertilizers into water courses. Most of the suspended sediment settles out, but colloidal-size material does not. Both suspended sediment and colloidal matter cause turbidity in surface water, making the water less attractive, inhibiting the growth of oxygen-producing algae, and possibly damaging marine life.

There are a number of problems involved in identifying the significance of nondegradable wastes. Each type of receiving water has different characteristics with the result that the same waste has different impacts on water quality. The concentrations of pollutant involved, especially concerning chemical substances, are often on the order of a few parts per billion; the extent to which continued low-level exposures to such concentrations are harmful to plant and animal life, and the extent to which there are synergistic reactions still is not well understood.

Degradable Wastes. The most damaging source of degradable waste is industrial discharges; the most common source is probably domestic sewage. Organic waste, which is highly unstable, can be converted to stable inorganic material such as bicarbonates, nitrates, sulphates, and phosphates by bacteria and other organisms in bodies of water. If the receiving waters are not overloaded with wastes, the process (usually referred to as self-purification) proceeds aerobically through the action of bacteria using free oxygen. If the receiving waters become overloaded with degradable wastes the process proceeds anaerobically through the action of bacteria not requiring free oxygen. The anaerobic process produces hydrogen sulfide and other unpleasant gases. Both processes are duplicated in conventional waste treatment plants, which merely accelerate the reactions that take place in natural waters.

Bacteria are always present in natural waters, and most types are harmless to man. Some bacteria arise from waste discharges of warm-blooded animals. The measure of such waste content, called the coliform count, is used as a proxy for the real bacterial concern —those bacteria capable of causing typhoid, dysentery, hepatitis, and cholera. Most bacteria are considered degradable, since they usually die quite quickly after leaving the body.

Thermal waste is usually classified as degradable because heat is readily dissipated in receiving waters, either by evaporation through surface water or conduction in groundwater. The principal sources of thermal pollution are the generation of electrical energy (including nuclear energy), and cooling operations in the petroleum refining, iron and steel industries, and other industries. The thermal barrier caused by a rise in temperature of more than 4°F. in receiving waters may have the effect of sterilizing or killing fish and other marine life.

The quality of a waste discharge can be measured through biological oxygen demand (BOD), chemical oxygen demand (COD), alkalinity or acidity, electrical conductivity, or turbidity. The most common measure of organic waste load is BOD, which indicates the quantity of oxygen used in the decomposition of the waste. While the level of dissolved oxygen (DO) is certainly not the only parameter of water quality,[2] it is common in empirical studies to use DO levels as indices of threshold conditions necessary for various water activities. For example, we may postulate that at least 3 milligrams per liter (mg/l) of oxygen are necessary to eliminate offensive odors and therefore to allow boating, 4 mg/l for sport fishing, and 5 mg/l for swimming. Thus, for pragmatic reasons, DO levels are often used as a first estimate of the level of water quality.

The amount of oxygen and its rate of use are functions of the type and quantity of waste, and the chemical characteristics of the receiving waters. For example, toxic materials may reduce the rate of decomposition by inhibiting bacterial action. At higher water temperatures the oxygen saturation of water is relatively low, bacterial action is increased, and biochemical oxygen demand increases as wastes are degraded more rapidly. If the imbalance between available oxygen and oxygen demand becomes too great, decomposition may become anaerobic. On the other hand, certain factors restore dissolved oxygen to the water. This reaeration process depends on such factors as velocity of stream flow, area of air-water interface, and photosynthesis.

The effects of discharged wastes are interrelated in many complicated ways. For example, water turbidity affects the photosynthetic production of algae, which requires sunlight. Thus, waterways with high turbidity usually do not have the odor and water taste problems that accompany prolific algae growth. When efforts are made to reduce turbidity, and increase the clarity of the water, algae production and resulting odor and taste problems may increase dramatically.

301. The Problem of the Oceans

While most of this chapter is devoted to the economics of the comparatively simple problem of fresh water pollution, the more difficult and potentially much more serious problem of pollution of the oceans must be mentioned at least in passing. The famed oceanographer and undersea explorer, Captain Jacques Cousteau, has estimated that the damage done to the oceans in the past 20 years is somewhere between 30 percent and 50 percent of that which would be required for most life in the oceans of the world to die under the stresses of pollution, just as Lake Erie has died. He claims that the amount of life in the oceans is decreasing rapidly, with the decrease first noticeable in 1968. In 1970 the world catch of fish dropped for the first time in two decades.[3]

We have given so little thought and care to the oceans because people have assumed that their immensity protected them; man could do nothing to disturb so gigantic a force. While scientists still feel that the oceans can absorb a good deal of waste without irreparable damage to ecosystems or to humans, it is now recognized that we don't have to pollute all 140 million square miles of ocean before we suffer unacceptable consequences—we only have to pollute a good part of the 14 million square miles closest to shore.

The sea as we know, is shaped like a deep bowl. However, waste that we throw into the sea does not spread out evenly to be diluted, nor does it necessarily come to rest in the ocean's deepest parts. Rather, most pollution remains on the continental shelf, which occupies only about 10 percent of the surface of the sea, but where 90 percent of the fish which are consumed are spawned, raised, and caught. The shelf extends out from land for anywhere from two miles to hundreds of miles, and in general follows the contours of the continents themselves.

It is from rivers that most pollutants reach the continental shelf, although harborside industries add their share of effluents, heated water comes from power plants, ships at sea contribute sewage and garbage and trash, and both ships and tankers contribute oil and sludge. It is estimated that a million tons of oil are spilled accidentally into the world's oceans each year, and more is added from offshore drilling and pumping rigs, from onshore accidents, and from the deliberate dumping of oil residues at sea by oil transports.

Most pollution that reaches the deepest parts of the ocean probably comes via wind and rain rather than from polluted rivers. The atmosphere deposits in the oceans an estimated 200,000 tons of lead and a million tons of oil from engine exhausts each year, and as much as 5,000 tons of mercury, which comes primarily from fuel consumption but is also used in the papermaking process and released into the atmosphere when the paper is burned. Probably half the pesticides in the sea have come from the air where they are carried long distances by high altitude winds. Cousteau estimates that 25 percent of all the DDT compounds so far produced have already been absorbed by the sea.

When people impose sudden loads of chemical compounds on an ecosystem, the system may be drastically altered. For example, human sewage will decompose in the ocean if given enough time, because salt is lethal to the bacteria in these wastes. But when receiving waters become overloaded with sewage, some of the bacteria survive long enough to be taken up by shellfish, which feed by straining water through their systems. Once the bacteria are inside the shellfish they can thrive, protected from the salt. Infectious hepatitis germs collect there in numbers sufficient to make the shellfish a hazard for man to eat. Shellfish have been found to concentrate viruses such as polio up to 60 times the proportion of viruses in the surrounding water.

Even where the amount of pollutant in the ocean is minimal in terms of the volume of water, three things tend to increase the level of pollutants reaching the food chain. The first is the capacity of marine organisms to store pollutants, which are passed along the food chain. The second is the location of so much pollution on the continental shelf. The third is the longevity of many pollutants, for example DDT, which has a half-life of about 15 years.

There may also be dramatic synergistic (or multiplicative) effects of these wastes on the environment. Nickel is a relatively nontoxic metal, but put into water with copper effluent it multiplies the toxicity of the copper by a factor of 10. If you add iron and zinc and sewage and heated water from thermal plants there is a synergistic mix which reacts to itself as well as to the chemistry of the sea, with ultimate effects that can only be guessed at.

302. Fresh Water Uses

The aesthetic enjoyment of bodies of fresh water (and of that portion of the oceans immediately adjacent to land) is primarily a matter of visual perception. At a minimum this means that the water must be free from obnoxious floating or suspended substances, particularly domestic sewage and industrial waste such as oil.

For recreational activities which require water contact, the water must not only be aesthetically pleasing but must contain no substances which are toxic upon ingestion or irritating to the skin or eyes. Also, the water must be relatively free of pathogenic organisms. Most efforts to measure water quality for recreation purposes have centered on this last condition, and a wide range of standards have been invoked, ranging from 50 to 3000 bacteria per 100 ml. Such standards do not appear to be based on epidemiological evidence of a direct relationship between contact with contaminated water and bacterial infections. In one study, McKee and Wolf concluded that the risk to health from bathing in sewage-polluted sea water was negligible. If any risk exists, it is probably associated with chance contact with solid lumps of infected fecal matter.[4] A greater possibility of infection exists in fresh water than in saline water, but the level of probability involved is almost certainly less than has been commonly assumed. For the most part, nondegradable wastes have little impact on the aesthetic enjoyment or recreational use of water unless the substances are toxic, or suspended sediment exists in concentrations high enough to impart an unattractive color to the water.

On the other hand, the value of water as a habitat for aquatic life is reduced or destroyed by some waste discharges which do not necessarily render the water aesthetically displeasing to human beings. The effects of these discharges on fish and other aquatic life are difficult to predict because they vary with the physical and chemical composition of the water. Higher temperatures, excessive acidity or alkalinity, and a low concentration of dissolved oxygen can increase the sensitivity of fish to toxic substances or can themselves result in fish kills. Conversely, certain combinations of salts can neutralize each other whereas acting independently they would be harmful.

Perhaps the most important factor in determining the tolerance of aquatic life to a pollutant is the time-concentration relationship. While a single short-term exposure to a high concentration of pollutant may show no damaging effects, repeated exposures to the same concentration or continuous exposure to a much lower concentration may result in death. On the other hand, with gradual exposure many organisms can develop tolerance to concentrations that would otherwise be toxic.[5]

The amount and character of treatment which is necessary before water can be used for domestic purposes is related to the quality of the water intake. Water which contains organic substances from domestic sewage must be treated with large amounts of chlorine or other disinfectants to kill the bacteria. If the water is corrosive, saline, hard, or contains quantities of substances like iron or manganese, further special treatment may be needed. There is little empirical evidence on which to base a limiting standard for drinking water with respect to total dissolved solids. The standard of 500 milligrams per litre (500 parts per million) is often used, but with no apparent epidemiological basis.

The effect of water quality on irrigation use cannot be defined in terms of a single water quality variable.[6] The extent to which crop yields are reduced by water quality deterioration is a function of the type of crop, type of soil, extent of drainage, the type of salts already in the soil, and other factors. Even changes in the temperature of applied water may improve or reduce crop yields.

The range of water quality used in industrial operations is very wide, depending in part on whether the water is used for processing, as boiler feedwater, for cooling, or for sanitary purposes. Feedwater for high temperature boilers requires the highest quality standards of any industrial use. Next to boiler feedwater, the quality standards for process water which comes into contact with foodstuff is the most demanding. At the other extreme, cooling water can be almost any level of quality with respect to total dissolved solids and even dissolved oxygen levels. With proper treatment it is quite possible to use sea water for cooling purposes.

303. Prediction of Dissolved Oxygen Levels

The rate of biochemical oxygen demand combined with the rate at which oxygen is restored determines the level of dissolved oxygen in a waterway. In flowing water the combined effect of a degradable waste discharge and reaeration in the stream produces a decrease and then an increase in DO as the waste moves downstream. Factors that reduce the rate of BOD lengthen and decrease this "oxygen sag" of DO in the water, while factors that accelerate BOD have the opposite effect. The oxygen sag is also affected by the rate of reaeration, which itself depends on water characteristics, photosynthetic oxygen production, and many other factors.

Despite these complexities, engineers have been able to develop models which interrelate the many variables involved and predict (given the nature and spacing of waste loads), the levels of dissolved oxygen, the temperature, and other water quality characteristics for different zones of a body of water. Such models provide a basis for economic optimization studies of water quality management systems, and determine, for example, the proper capacity and design for a waste treatment plant. Basic equations describing DO behavior can be elaborated to handle complex situations, including multiple points of waste discharge, different kinds of waste, and differing oxygen saturation along the length of the stream.[7]

There are a number of complexities which can complicate the calculation of DO levels. Tracing the decomposition of organic waste in an estuary is made difficult by tidal actions, the large air-water interfaces, and the complex hydraulic characteristics of the body of water. Lakes present problems because their more stationary water is more susceptible to nutrient buildup (eutrophication), a process which is accelerated by organic wastes.

Nutrient enrichment tends to be accompanied by low DO levels, but the complex circulation patterns and temperature inversions combine to make the exact effect of organic waste discharges on water quality difficult to predict.

A final complication is that BOD proceeds in two stages. When an organic waste is discharged into a stream where DO levels are high, there is an immediate decrease in DO level as organic wastes are degraded by bacterial action. Thereafter, DO levels tend to recover. After five to seven days, a second stage BOD occurs as the nitrogen in the organic wastes is converted first to nitrite and then to nitrate by aeorbic nitrifying bacteria. The second stage is more diffuse and easier to predict than the first.

Lakes, reservoirs, and estuaries, like flowing streams, are subject to deposits of sludge banks, which if suddenly dispersed (for example through dredging) produce a "shock load" of oxygen demand. Stratification, the formation in essentially stationary bodies of water of thermal layers which prevent vertical mixing of water, complicates the analysis of DO levels in lakes and reservoirs. Plant nutrients take on special significance here because they tend to accumulate over time in relatively stationary bodies of water and contribute to the depletion of oxygen in the lower levels. In a tidal estuary, areas of water with low DO levels may be moved back and forth within the estuary for long periods of time rather than being dispersed into the ocean or undergoing vertical mixing with higher DO areas of water.

ECONOMIC APPROACHES TO WATER POLLUTION ABATEMENT

Given the complexities involved in calculating water quality standards and water pollution effects, it is helpful when taking an economic approach to water pollution optimization to attack the complex issue by first assuming away some of the complexities, and then solving the simplified problem that remains. The value of this exercise is not in the resulting answer, which is at best a first approximation, but in isolating the assumptions (and thus the knowledge required) to formulate a more realistic problem. Water pollution is more amenable than most other environmental problems to such an approach because so many of the variables involved are—at least potentially—capable of being quantified.

304. Economic Approaches to Water Pollution Assuming Complete Knowledge

Consider the problem faced by the citizens of the town of North Lake Tahoe, located on Lake Tahoe in Nevada.[8] The waters of Lake Tahoe were until a few years ago remarkably pure, but they have been increasingly polluted because the inflow of sewage effluent from the town slightly exceeds the capacity of the lake to cleanse itself. So far only the one pollutant (sewage) is involved, and it exists in approximately equal proportions from each citizen of the town.

The citizens of the town have several options or alternatives open to them in preserving the lake's water for drinking purposes. Each family can add chlorine pills to its drinking water at a cost of $6 per person per year (or $3 per half year, or $4.50 for 9 months). Or, engineering studies indicate that a water treatment plant can be built to remove all noxious wastes from the drinking water at a cost of $5 per citizen per year. One that will remove 75 percent of the wastes can be built for $3.50 per citizen per year, or one that will remove half the wastes for $2 per citizen per year. A further study indicates that within ten years Lake Tahoe will be unfit even for swimming unless a sewage treatment plant is also built, and this will cost each citizen $3.00 per year. The sewage treatment plant will solve only the swimming problem; the drinking water problem is solved only through the construction of the water treatment plant, or by the addition of chlorine pills to drinking water.

A final medical study indicates that if nothing is done, water pollution will result in one day's mild sickness per citizen per year. If 50 percent of the wastes are removed, sickness falls to three days per ten years (because the illness caused by sewage goes up faster than does concentration of pollutant). Similarly a plant that is 75 percent effective cuts sickness rates from water pollution to one day per citizen in ten years.

A quick survey of a sample of Lake Tahoe citizens indicates that each citizen is prepared to pay $2.50 per year for the privilege of swimming in the lake. A second survey shows that the average citizen will pay $10 to avoid a day's sickness. Thus, a treatment plant that is 100 percent effective is worth $10 a year to each citizen.

It is now possible to put together a simple matrix (see Table 3-1), in which we specify the alternatives open to the citizens of the

TABLE 3-1

Cost-Benefit Matrix For Water Pollution Abatement of Lake Tahoe

	Add Chlorine to Drinking Water (% of water treated)				Build Water-Treatment Plant (% of water treated)				Build Sewage Plant	Build Water-Treatment Plant and Sewage Plant (% of water treated)				Do Nothing At All
1. Policy	100 a	75 b	50 c	0 d	100 e	75 f	50 g	0 h		100 i	75 j	50 k	0 l	m
2. Damage Avoided (per citizen)	$10.00	$9.00	$7.00	$0	$10.00	$9.00	$7.00	$0	$2.50	$12.50	$11.50	$9.50	$2.50	$0
		(health)				(health)			(swimming)		(health and swimming)			
3. Cost of Avoiding Damage (per citizen)	6.00	4.50	3.00	0	5.00	3.50	2.00	0	3.00	8.00	6.50	5.00	3.00	0
4. Net Benefit (#2 − #3)	4.00	4.50	4.00	0	5.00	5.50	5.00	0	−$.50	4.50	5.00	4.50	−$.50	0
5. Cost of Damage Not Avoided (per citizen)	2.50	3.50	5.50	12.50	2.50	3.50	5.50	12.50	$10.00	0	1.00	3.00	10.00	12.50
	(health)		(health and swimming)		(health)		(health and swimming)		(health)		(health)			
6. Total Costs Incurred (#3 + #5)	8.50	8.00	8.50	12.50	7.50	7.00	7.50	12.50	13.00	8.00	7.50	8.00	13.00	12.50

town. The alternatives are indicated horizontally at the top of the matrix, and calculations are made vertically. Citizens can make a decision based either on maximizing net benefit per citizen (No. 4), which is damage avoided less the cost of avoiding it, or by minimizing the sum of damage-avoidance costs plus value of damage not avoided (No. 6). The two solutions will always prove to be equivalent. The citizens decide to build a water-treatment plant to remove 75 percent of the sewage for $3.50 per year per citizen. They decide also not to build a sewage plant, and not to undertake any addition of chlorine tablets to drinking water.

On checking the calculations it is obvious that the answer obtained depends completely on the numbers assumed, and on their accuracy. If the cost of removing all the sewage from drinking water were actually $4.40 per citizen per year instead of $5.00, the net benefit from this strategy would be $5.60 instead of $5.00, the total cost incurred would be $6.90 instead of $7.50, and 100 percent abatement rather than 75 percent would be adopted. A mis-estimate from $5.00 to $4.40 could arise through engineering error, or the cost of building a plant might decline by the requisite 12 percent as water treatment technology improved. Thus, the choice of preferred strategy is highly sensitive to the estimation of the costs of water treatment. Similarly, a small increase (from $2.50 to $3.00 or more) in the amount that a citizen is prepared to pay for the privilege of swimming in the lake would mean that a sewage treatment plant would become part of the optimal solution.

Reaching a solution in the matrix was possible only because some very simplistic assumptions were made. It was assumed that the characteristics of the "average" citizen of North Lake Tahoe were known, i.e., the quantity of pollutant he contributed, the value he placed on a day's illness, and the value he placed on recreational swimming. We also assumed that the damages avoided by alternative solutions were known, and were measurable in dollars. We ignored the technical problem of interrelationships among pollutants by assuming that sewage was the only pollutant. Finally, we considered only the citizens of North Lake Tahoe, and ignored how sewage pollution affected residents on the south shore of the lake, and how the quantity of pollution originating on the south shore might vary depending on what was done by the citizens of North Lake Tahoe. Each of these assumptions is important; to assume away any one distorts the entire analysis.[9]

Some of this information is lacking because it is not available (although obtainable), and some because it is not obtainable given present measurement techniques. The technical problem of interrelationships among pollutants is certainly solvable, although complicated; ecologists claim that accurate estimates of the interrelationships among pollutants, and good predictions of the damage avoided by removing any combination of pollutants, can be made. As indicated earlier, the technical problem of measuring the value of avoiding a day's sickness is complex; people simply do not *know* what they might be willing to pay to avoid the imprecisely stated effects of pollution. If they did know, they might overstate their answer in the hope that someone else, "the government," would be more willing to correct the situation. If a respondent felt he would be required to contribute funds for pollution abatement equal to the damage he had suffered he might avoid the expense by understating his own damage in the hope that others would contribute enough to implement abatement anyway.

Nor would a simple voting system invariably work to insure an economic amount of pollution abatement. Thirty of the residents of North Lake Tahoe might incur health and swimming damage of $12.50 each, and twenty others damages of $17.50 each. A vote on an abatement proposal costing $13.50 per citizen per year then would fail, even though total community damages of $725 could have been avoided at a community cost of $675.

The most difficult factor to measure may be recreation values such as swimming rights. However, aesthetic enjoyment and freedom from health hazards are one of the largest components of water pollution problems and cannot be ignored.

305. Economic Approaches to Water Pollution Assuming Imperfect Knowledge

The opportunities for economic analysis to be used in choosing an optimal pollution-abatement policy from among all possible alternatives in a world where people differ and information is incomplete are somewhat limited. Once we discard the assumption of complete information, it is difficult and misleading to draw up a matrix showing an accurate dollar value of benefits and costs of all possible abatement strategies.

But difficult or not, something usually has to be attempted. Like most problems, the pollution of Lake Tahoe refuses to go away

TABLE 3-2

Town Commission Decision Matrix for Water Pollution Abatement of Lake Tahoe

	Add Chlorine to All Drinking Water At Source	Build Water-Treatment Plant With Tax Funds (% of water treated) 100 75 50 0	Build Sewage Plant With Tax Funds	Build Water Treatment Plant and Sewage Plant With Tax Funds (% of water treated) 100 75 50 0	Do Nothing At All
1. Government Policy					
2. Damage Avoided (per citizen)	Marked improvement in drinking water quality but deterioration in taste. Health benefits at least $10.00 per annum, probably much more	Marked improvement in drinking water quality and some improvement in taste. Health benefits up to $10.00 per annum, perhaps more, and some aesthetic benefits which are difficult to measure.	Marked improvement in swimming, value is hard to measure.	Marked improvement in drinking water quality and in swimming, health benefits of up to $10.00 per annum or more plus aesthetic benefits plus value of swimming improvement which is hard to measure.	None
3. Cost of Avoiding Damage (per citizen)	$4.00	$5.00 $3.50 $2.00 $0	$3.00	$8.00 $6.50 $5.00 $3.00	$0
4. Net Benefit (#2 − #3)	uncertain	uncertain	uncertain	uncertain	$0
5. Cost of Damage Not Avoided (per citizen)	Present danger to swimming and health hazard from swimming remain, value hard to measure.	Present danger to swimming and health hazard from swimming remain, value hard to measure. Some health hazard if water treatment less than 100 percent.	Health hazard from drinking water at least $10.00 per annum, probably more.	All damage avoided if 100 percent of water is treated, some health hazard from drinking water if treatment less than 100 percent.	All health hazards from drinking water and damage to swimming remain, value of $10.00 per annum or more.
6. Total Costs Incurred (#3 + #5)	$4.00 per citizen plus damages to health and swimming.	Up to $5.00 per citizen plus damages to health and swimming.	$3.00 per citizen plus health hazard from drinking water.	$8.00 per citizen, or $3.00 to $6.50 per citizen plus damages to health and swimming.	Health hazards from drinking water and damages to swimming.

by itself. There is increasing citizen pressure for government to do something, so the North Lake Tahoe Town Council appoints a Commission to look into the facts, and they prepare their own revised matrix. It is apparent that some people are getting sick from drinking untreated water; that swimming in the lake is not as pleasant as it used to be, and is getting worse; that some residents have already installed individual water-treatment systems at considerable expense and that other people have sold their lakeside cottages but have been unable to realize enough from their sale to buy equivalent properties elsewhere. Only a small number of people are spending any money to abate water pollution damage on their own, and the damage suffered by those who do nothing is very difficult to quantify except in the most general terms.

Town engineers are able to tell the Commission what the per capita costs of avoiding damage are under the varying alternatives available. The cost of adding chlorine at the source is somewhat lower than it is when added individually. The cost of building a water-treatment plant and/or a sewage plant are the same as in the earlier example. The net benefit per capita (No. 4 in the matrix of Table 3-2) becomes uncertain or unknown for four of the five alternatives under a situation of incomplete knowledge. The total costs incurred are still the per capita costs of avoiding damage from No. 3, plus the residual damage as indicated in No. 5. Even in this example, the Commission has not considered water use problems other than fitness for swimming and drinking—for example quality of fishing, the growth of algae in the lake, or the buildup of sludge on the lake bottom. The citizens of North Lake Tahoe have also ignored (quite properly from their point of view) the impact of lake pollution costs on others, although it is obvious that some cost is imposed on everyone who lives on the circumference of the lake, and on all those who use it periodically for recreational purposes.

Nevertheless, the Commission puts together a Town Decision Matrix for Water Pollution Abatement, which, if less precise than the original cost-benefit matrix, is at least a systematic way of looking at the problem, particularly of weighing what is achieved (in No. 2) and what is not achieved (in No. 5), under each alternative. Such a decision matrix might help if the issue were put to voters for a decision (for example on raising taxes for water pollution abatement), but the voting scheme would likely prove unfeasible. As a minimum it would be necessary to have several elimination votes. Otherwise the policy of "Do Nothing At All" might win with a plurality but not a majority, while a majority of people would have preferred to compromise on a policy such as a moderate degree of water treatment which, although not their favorite, would have been preferred to the option of "Do Nothing."

Even solving the method of voting does not answer the question of who should vote—permanent residents of North Lake Tahoe certainly, and probably temporary residents of the town who are also land owners or property owners. But should people on the south shore who also have a stake in the quality of the water (and who incidentally are located in California rather than in Nevada) also be allowed to vote? It could be argued that everyone in northern Nevada and many people in northern California have some interest in the outcome, and should be allowed some voice in it. These problems, and that of citizen indifference, mean that voting is at best impractical, and at worst useless as a solution to this real-world pollution problem.

Given continued citizen pressure to do something about the problem, the Commission will choose some solution, based not on economics but on some sort of political reality. If the Commission chooses one of the policies with the lowest cost of avoiding damage (doing nothing, building only a sewage plant, or building a water treatment plant for only a small portion of the water taken from the lake), citizens who are concerned about pollution will see the lake and their drinking water continue to deteriorate, and will be unhappy. If the Commission chooses the most expensive policy (that of building a treatment plant for all the water plus a sewage plant), taxes will rise considerably and citizens who object less to pollution than to high taxes will be unhappy. The Commission will almost certainly choose some intermediate solution such as 50 percent water treatment, arguing that it is wiser to start with intermediate treatment and see how citizens react to the new pollution levels and tax load, and add on further water treatment and/or a sewage plant in the future if this seems appropriate. Such a political decision can be objected to as inferior, but only in comparison with an economic decision based on perfect information which of course we do not have. Lacking perfect information, the Commission's decision cannot be proven economically superior or inferior to the other available alternatives.

One outcome of a local-Commission type of decision is the tendency to a pollution-zoning of different areas. Different cities and towns will differ significantly in their pollution levels and in their tax assessments for pollution control, and people will be induced to move until they find the combination of water quality and tax level that they can tolerate. After people have sorted themselves out in relation to geographic pollution levels, the amount of political complaint about environmental pollution can be expected to decline.

For a much more complex economic analysis of a water pollution problem, consider the example which follows of the Delaware River Estuary.

THE DELAWARE RIVER ESTUARY STUDY: A CASE STUDY

306. General Background to the Delaware River Study

During the late 1950's, several state and interstate water pollution control agencies and the City of Philadelphia became concerned with the severe pollution of the Delaware Estuary. They requested the Public Health Service's Division of Water Supply and Pollution Control, now the Federal Water Pollution Control Administration, to develop a program for water pollution control in the Delaware Estuary. The Delaware Estuary Comprehensive Study (DECS) was undertaken in late 1961 in cooperation with the State regulatory agencies of New Jersey, Pennsylvania, and Delaware, the Delaware River Basin Commission, the City of Philadelphia, and a number of other interested parties. The study area encompassed the Delaware Estuary from Trenton, New Jersey to Liston Point, Delaware, including the estuarine reaches of its tributaries.

The decision-making body which resulted, the Delaware River Basin Commission, was at that time the only interstate-federal compact agency in the United States. The objective of the Commission was to devise, based on the study, a multipurpose water resources plan to upgrade the river and preserve it for a variety of uses; to bring the greatest benefits and produce the most efficient service in the public welfare.

The Commission in its study was thus faced with a magnified version of our simplified water quality problem, with some of the political jurisdiction issues stripped away but complicated by the necessity of dropping many of the simplifying assumptions, and of quantifying some of the values ignored in the example. The sections that follow discuss only a small portion of the total study, but indicate how the problems involved can be approached, given sufficient time and resources.

Using computers, the study devised a mathematical formulation of the entire Delaware estuary, and the researchers were able in a short time period to collect and analyze technical data on water quality that earlier would have required years.[10] The river mathematical model, an untested approach until the DECS application, has now become a proven mechanism for application to previously unsolvable water pollution problems.

During 1966 about 28,000 people were employed by the firms designated as substantial waste dischargers in the Delaware River Estuary area. For the 20 major industrial waste sources, the estimated dollar value of output was about $4 million. The total carbonaceous oxygen demanding waste load discharged into the estuary during 1966 was 1,400,000 pounds per day. About 65 percent of this discharge was from municipal discharges and 35 percent from direct industrial discharges. There was an oxygen demand of about 200,000 pounds per day exerted by bottom deposits of sludge and mud, which were the result of material discharged from storm-water overflows, from municipal and industrial waste effluents, and from dredging operations.

The vast majority of municipal waste effluent flows were discharged without disinfection and consequently contained large concentrations of coliform bacteria. During 1966 all municipal sources along the estuary gave at least primary waste treatment (about 30 percent removal of oxygen demanding load), and some waste treatment was as high as a 90 percent removal level. The amount of industrial waste reduction along the estuary ranged from none to a 95 percent removal of "raw" load. During 1966 the average removal of waste discharges along the estuary was about 50 percent of the raw load.

Population projections in the study area indicate increases of 30 percent between 1960 and 1975, and of 135 percent between 1960 and 2010. It was estimated that 1966 raw waste loads would increase by 2.3 times by 1975, and by 5 times by 2010. Industrial raw waste loads were expected to double from 1966 to 1975, and to increase by more than six times by 2010. Overall the total municipal and industrial waste load prior to treatment was expected to double between 1966 and 1975, and to increase by about 5½ times by 2010.

307. Water Quality Goals[11]

The procedure used in establishing water use and water quality objectives for the Delaware River Estuary was to investigate all feasible water uses, to determine water quality criteria sufficient to guarantee those uses, and to assign water quality goals to the various sections of the estuary according to the uses designated. Meetings of the Water Use Advisory Committee (WUAC) were held to elicit community feelings on possible swimming areas, on desirable fishing locations, on withdrawal of water from the estuary, and on intentions of potential industrial water-users. Based on the work of WUAC, the thousands of possible combinations of uses versus location were reduced to five sets of possible water use and associated water quality objectives. The five objective sets ranged from maximum feasible enhancement of the river under present technology to maintenance of existing levels of water use and water quality. For each set of objectives the costs were estimated and the benefits, where possible, were quantitatively evaluated. It was not required that the final objective be any one of the individual sets but could be composed of various features from each of the objective sets.[12] The five water use/water quality sets were as follows:

Objective Set I   (OS I) represented the greatest increase in water quality levels among all the objective sets. Water contact recreation was anticipated in the upper and lower reaches of the estuary, with sport and commercial fishing envisioned in other areas. A minimum daily average DO level of 6.0 mg/liter was included for anadromous fish passage during appropriate periods. Fresh water inflow controls were proposed to overcome high chloride concentrations in several specific locations. OS I required 92–98 percent removal of all carbonaceous waste sources, plus instream aeration. An estuary-wide residual of 100,000 pounds per day of oxygen demanding wastes was permitted. There was considerable uncertainty as to the ability to achieve these reductions over the entire estuary. OS I required large-scale utilization of advanced waste treatment and reduction processes which had doubtful technical feasibility in 1966.

Objective Set II (OS II) anticipated a reduction in the area set aside for water contact recreation in OS I, a reduction in minimum DO levels with a concomitant reduction in sport and commercial fishing, and reduced chloride control as compared with OS I. OS II required removal of approximately 90 percent of the existing waste load with an estuary-wide residual of 200,000 pounds per day of oxygen demanding wastes permitted.

Objective Set III (OS III) was identical in all respects to OS II except that no DO criteria for anadromous fish passage were imposed, thus a further decrease in sport and commercial fishing potential was anticipated. Also, water quality standards at points of municipal water supply were reduced from those anticipated in OS I and OS II. OS III required removal of about 75 percent of the existing waste load with a residual load of about 500,000 pounds per day allowed.

Objective Set IV (OS IV) represented a slight increase over then-existing levels in water contact recreation and fishing in the lower reaches of the estuary. Generally, quality requirements were increased slightly over 1964 conditions in OS V, representing a minimally enhanced environment. OS IV called for about a 25 percent removal of existing waste loads with a residual load of 650,000 pounds per day allowed.

Objective Set V (OS V) represented a maintenance of 1964 water quality conditions, and was intended to prevent any further deterioration of water quality levels from those then in existence. OS V permitted a dumping of about 950,000 pounds per day of oxygen-demanding wastes along the reaches of the estuary.

308. Alternative Approaches To Water Quality Objectives

The methods considered for improvement of water quality in the river included limiting effluent discharge to the estuary by requiring reduction of wastes before discharge; piping of wastes to other places, where the discharges would have a reduced economic effect; regulation of stream flow; removal of benthic sludge deposits; instream aeration; and control of storm water discharges. It was concluded that a comprehensive program might incorporate several of these possibilities, but would have the greatest assurance of success if it depended primarily on reduction of waste at the source.

There are many ways of controlling the discharge of waste to achieve a specified water quality objective. The problem is to choose a system which balances the apparent equity of the solution to the individual waste discharger, the economic cost to the region, and the means of administering the water quality control program. The DECS used an economic model of dissolved oxygen conditions to investigate four alternate control programs for achieving desired DO levels in the estuary. The DO model is a good example of the sort of analytical technique available, and is discussed briefly below.[13]

The DECS computer model segmented the watercourse into 30 sections of 10,000 to 21,000 feet in length each; the segmentation permitted the prediction of effects of a change in waste loads in one section upon all other affected sections. Superimposed on this physical model of the estuary is a cost optimizing economic model (a linear programming formulation) which considers imputs such as: (1) location of a waste source with respect to the DO profile of the waterway; (2) the relative cost of removing waste at each source; (3) the maximum quantity of waste that can be removed at each source; and (4) the proximity of one waste discharge point to existing points. With cost data collected by sampling and survey methods, the study analyzed four programs for achieving alternative DO objectives in the estuary. The control programs considered were:

Cost Minimization (CM), which uses a mathematical programming solution to obtain the minimum total cost of waste treatment yielding the desired DO level for all individual polluters in the region.

The CM solution results in differing levels of treatment at different pollution points because treatment is concentrated at those points where the critical oxygen sag can be reduced most inexpensively.

Uniform Treatment (UT), which requires all waste dischargers to reduce their waste loads by the same percentage, with the percentage chosen being the minimum needed to accomplish desired DO levels.

Single Effluent Charge (SEC), which requires each waste discharger in the estuary to pay a uniform price per unit of oxygen demanding material discharged. The solution estimates the minimum single charge which will induce sufficient reduction in waste discharge to achieve desired DO standards.

Zoned Effluent Charge (ZEC), which uses a uniform effluent charge in each of a number of zones, instead of a uniform charge over all reaches of the estuary.[14]

The economic costs associated with the four programs are shown in Table 3-3 for two levels of water quality. The 3-4 ppm standard is the one that estuary authorities considered the maximum practically attainable in the estuary.

TABLE 3-3
Costs Associated with Various Control Programs

D.O. Objective	Cost Minimization (CM)	Uniform Treatment (UT)	Single Effluent Charge (SEC)	Zoned Effluent Charge (ZEC)
	(all figures in millions of dollars per year)			
2 ppm	1.6	5.0	2.4	2.4
3-4 ppm	7.0	20.0	12.0	8.6

While the CM solution is the most efficient in each case since it programs waste discharges at each point specifically in relation to the cost of improving quality, this comes at the cost of highly detailed information on treatment costs at each point, and an extremely inequitable distribution of costs.[15] The CM solution is closely approximated by the ZEC solution at the higher DO objective level. In effect ZEC "credits" upstream dischargers with the waste degradation that takes place in the stream, a necessary condition for full efficiency when effluent charges are used to achieve a minimum standard at a critical point in a waterway. The ZEC solution does not achieve the full efficiency of the CM solution because the basis for the "credit" is too broad.

An effluent charge of about 10 cents per pound of BOD would be needed for the ZEC solution; using this charge the administrative agency would collect $7 million per year in rent on the assimilative capacity of the watercourse.[16] For industry and municipalities, this is about the same as the cost of treatment only under a uniform treatment program. The study concluded that a charge at that level would not cause major regional economic readjustments such as the closing of industrial plants in the study area. Perhaps the great advantage of the ZEC or SEC approaches is that they require much less in the way of information and analytical refinement than does the CM solution.

309. Costs of Alternate Programs

The nature and amount of the current waste load in the estuary was determined by technical studies of rates of decay, reaeration, and other stream characteristics. Data on abatement costs and future waste load estimates was collected from existing municipal and industrial dischargers. Table 3-4 shows the estimated cost of achieving each objective set under three different cost-allocation formulas.

TABLE 3-4

Summary of Total Costs of Achieving Objective Sets 1, 2, 3, and 4 (Costs include cost of maintaining present (1964) conditions and reflect waste-load conditions projected for 1975-80). Flow at Trenton = 3,000 cfs

(million 1968 dollars)

	Uniform treatment			Zoned treatment			Cost minimization		
Objective set	Capital costs	O & M costs[1]	Total costs	Capital costs	O & M costs[1]	Total costs	Capital costs	O & M costs[1]	Total costs
1	180	280 (19.0)	460[2]	180	280 (19.0)	460[2]	180	280 (19.0)	460[2]
2	135	180 (12.0)	315[3]	105	145 (10.0)	250[3]	115	100 (7.0)	215[3]
3	75	80 (5.5)	155[3]	50	70 (4.5)	120[3]	50	35 (2.5)	85[3]
4	55	75 (5.0)	130	40	40 (2.5)	80	40	25 (1.5)	65

[1]*Operation and maintenance costs, discounted at 3 percent, twenty-year-time horizon; figures in parentheses are equivalent annual operation and maintenance costs in millions of dollars/year.*
[2]*High-rate secondary to tertiary (92-98 percent removal) for all waste sources of all programs. Includes in-stream aeration cost of $20 million.*
[3]*Includes $1–$2 million for either sludge removal or aeration to meet goals in river sections #3 and #4.*
Source: DECS p. 58.

An alternative to on-site waste reduction would be to pipe wastes out of the area of the estuary, presumably into the Atlantic Ocean. The obvious disadvantage of piping is that it merely moves the pollution problem from one location to another, perhaps externalizing pollution costs from one area to another. Table 3-5 shows the estimated cost of waste reduction versus piping for Objective Sets I through IV.[17]

TABLE 3-5

Capital Costs for Attainment of Objectives (millions of dollars) 1) By Piping of Wastes out of the Estuary; 2) By Reduction of Wastes at the Source

Obj. Set	Estimated[1] Diverted Flow (cfs)	1) Piping of Wastes Out of the Estuary			2) Waste Removal
		Piping	Chloride[2] Control	Total	
1	1200	125	40	165	180
2	1150	120	35	155	115
3	800	90	25	115	50
4	650	65	20	85	40

[1]It is assumed that industrial waste streams will be separated to allow cooling water to return to the stream.
[2]Estimated Capital Cost of additional storage necessary to counteract effects of diverted flow.
Source: DECS p. 68.

Rough estimates of the total cost of reaching the various DO objectives by mechanical aeration, including capital and operation and maintenance, are shown in Table 3-6. It should be noted that mechanical aeration meets DO objectives only, and additional expense would be necessary to meet other parameter objectives. Since large-scale in-stream aeration such as would be required for the Delaware has never been attempted except on a pilot-plant scale, considerable study would have to be devoted to the feasibility of the size of the system required. The cost estimates given thus are highly tentative. Also, it was anticipated that some problems might develop in interferences with navigation and recreation as well as in the creation of nuisance conditions, such as foaming.

TABLE 3-6
Estimated Total Cost to Reach DO Objective by
Mechanical Aeration

Objective Set	Cost (Millions of Dollars)
I	$70
II	40
III	12
IV	10

Economic projections showed that a substantial increase in waste production could be expected in the estuary area. To maintain any given objective set under increased waste loadings would increase program cost by an additional 5.0 to 7.5 millions of dollars per year from 1975 through 1985. No estimate of treatment costs after 1985 was attempted; it was felt that this would be misleading as other alternative methods of effluent treatment became feasible, as more efficient production processes became available, and as entirely new objective functions were undertaken.

310. Quantification of Benefits from Improved Water Quality

The next step undertaken by the DECS was to define and quantify the benefits of enhanced water quality in the Delaware Estuary. Quantification of benefits is a standard part of any engineering feasibility study. In this situation, a number of existing intangibles required that subjective value-judgments based on the estimated social satisfactions from improved water quality be made.[18]

Initially, data was collected from the major water-using industries along the estuary. DO and chloride levels were found to be the most important quality parameters to industrial water users. It was found that increased DO levels resulted in negative benefits (or increased costs) to water-using industries, primarily due to increased corrosion rates at the higher oxygen levels. The increase in cost ranged from $7 million for OS IV to $15 million for OS I. (This means that while humans seek higher DO levels, industry seeks lower levels). Chloride goals in OS II and OS III resulted in a benefit to industrial water users of almost $4 million per year, with the chloride goal in OS I producing an additional benefit of $2 million per year.

A study was then made to define and quantify the benefits that would accrue to the commercial fishing industry. Although the estuary itself does not support a commercial fish harvest of any size, its water quality does influence commercial fish production in adjacent areas.[19] In calculating benefits, a given species was considered to be beneficially influenced by improved water quality if it must depend on water within the study area for survival at some period in its life cycle. The estimated net commercial fishing benefits range from $3 million minimum to $5 million maximum for OS IV to $9 million minimum to $12 million maximum for OS I.

The next step of the DECS was to quantify the recreational benefits included in swimming, boating, and sport fishing. The benefits associated with other activities, such as picnicking and sightseeing, were seen as resulting from the improved aesthetic surrounding, but were considered to be non-quantifiable. The net dollar benefits that might accrue in the 1975-1980 period from increased recreational possibilities for each objective set were produced by: (1) estimating the total recreational demand in the Delaware Estuary region by applying national average participation rates to the region's projected population; (2) estimating the maximum capacity of the estuary under each of the objective sets; (3) estimating the part of the total demand expected to be filled by the estuary; and (4) applying monetary values to the estimated total participation demand in the estuary to arrive at total estimated benefits from recreation. The analyses indicated a tremendous latent recreational demand in the estuary region that to some extent could be satisfied by improved water quality. It was estimated that during the period 1975-1980 the increase in total demand for the whole region over the present demand would be about 43 million activity days per year, and by the year 2010 would increase by almost 100 million activity days per year.

The problems associated with estimating a figure for benefits from recreation are substantial. For example, suppose the report claims $110,000 in swimming recreational benefits for a community under OS III, but fails to recognize that the same swimming recreational benefits could be accomplished through an expenditure of $80,000 for a community pool. Which alternative measure of benefits should be accepted?

Some people would maintain that a public program which provides $1 million of benefits to one small segment of the population (recreational boaters), is inferior to a program that distributes $1 million of benefits to a broader sector of the population (swimming pools to poor inner city residents who cannot afford to travel to more distant parts of the estuary to swim). It is not clear whether cost-benefit analysis can be adapted to take into account the income distributive and benefit distributive effects of a recreation program, or for that matter of a sewage program. If such distributive effects cannot be quantified into the dollar units that are standard in cost-benefit analysis, how can such considerations be integrated into the planning process?

The estimated range of recreational benefits for each objective set is indicated in Table 3-7.[20] In this calculation, the benefits accruing to industry and the municipalities were seen as being small and as cancelling out because of the negative features of industrial water use. The ranges of recreational benefits were thus taken to be estimates of the total benefits from improved water quality in the estuary.

TABLE 3-7

Costs and Benefits of Water Quality Improvement in the Delaware Estuary Area[1]

(million dollars)

Objective set	Estimated total costs	Estimated recreation benefits	Estimated incremental cost		Estimated incremental benefits	
			minimum[2]	maximum[3]	minimum[2]	maximum[3]
1	460	160-350				
			245	145	20	30
2	215-315	140-320				
			130	160	10	10
3	85-155	130-310				
			20	25	10	30
4	65-130	120-280				

[1] All costs and benefits are present values calculated with 3 percent discount rate and twenty-year time horizon.
[2] Difference between adjacent minima.
[3] Difference between adjacent maxima.
Source: DECS data summarized in Kneese and Bower, Managing Water Quality: Economics, Technology and Instructions (Baltimore: Johns Hopkins Press, 1968), p. 233.

Table 3-7 indicates that OS IV appears to be justified, even when the lowest estimate of benefit is compared to the highest estimate of cost. The incremental costs suggest that going to OS III is marginal, but perhaps justifiable on the assumption that some of the more widely distributed benefits of water quality improvement may not have been taken into account. Clearly the incremental economic benefits of going to OS II or to OS I are outweighed by the incremental costs of doing so.

Note, however, that in addition to the benefits measured in the study there are numerous other uses that will be improved as a results of improved water quality. The water quality levels in any of the first four objective sets would reduce the rate of corrosion, delignification, and cavitation of piers, wharfs, bridge abutments, and boat engines and hulls. The quantity of debris, silt, oils, and grease that settle and block channels and cooling systems in boat engines would be reduced substantially. The dollar benefits attributable to these effects remain undefined.

Another important benefit of improved water quality is the improved aesthetic value of the river. Some of this benefit is included in the estimate of increased recreational value, but this estimate does not include the increase in value of property adjacent to the estuary, nor the enhanced value of parks and picnic areas adjacent to the watercourse.[21]

What is less apparent is that once the water quality reaches a threshold level at which several important recreational activities may occur, an additional increase in water quality may produce few if any new benefits. For example, once the bacterial standard for water contact recreation is such that swimming and water skiing can be authorized, no further benefit (except perhaps an unrecognized health benefit or some benefit from the existence of a "safety margin") results if bacterial levels are reduced further. In fact, improving DO levels above standard may produce negative benefits as well as increased costs because of the increase in oxygen-aided corrosion and similar effects.

The quantitative analyses also do not include the influence of secondary effects on the regional economy. For example, a unit of monetary benefit associated with commercial fishing use might be expected to generate at least an extra 15 percent in other benefits due to the interrelationship between the commercial fishing industry and the rest of the economy. This may occur in the form of increased wages, additional capital investment, or increased use of trades and services.

The use of a static analysis in the DECS conceals a number of additional factors: (a) the raw sewage load is growing at 8 percent a year; (6) undercapacity of abatement facilities can be corrected within 6 years if no preliminary planning has been done, or within 3 years if the necessary design work has been done on a contingency basis beforehand; and (c) the technology of waste treatment is evolving at a very rapid rate. How would these factors, incorporated into a dynamic rather than a static analysis, affect the conclusions arrived at by the DECS?

If a safety factor is felt necessary, should it be built into the objective set, or into the calculation of permissible waste loading in the objective set which we would like to achieve? Does the "location" of such a safety factor make any difference in terms of the ease of having the objective set adopted, the ease in resisting pressures from dischargers, or the avoidance of public disillusionment with pollution control and consequent loss of interest or overreaction? A safety factor was built into the objective set rather than in the permissible waste loadings in the DECS study to placate the large industrial and commercial representation on the advisory committees to the study.

311. What's Best for Philadelphia

The preceeding analysis has discussed the costs and benefits of water quality improvement for the whole of the Delaware Estuary area. However, voting on the various alternatives is done by the individual entities, primarily cities, which make up the Delaware Estuary compact, and it is enlightening to look at the political considerations facing these entities. The economic and political alternatives facing the City of Philadelphia were typical.

Under OS II, Philadelphia would be required to institute BOD removal of about 88 percent of its current raw sewage load at each of its three plants. OS III would require about 75 percent abatement at the current raw loading. The required percentage of BOD removal would rise as the raw load grew. The Philadelphia Water Department estimated that to build plants to allow the city to meet its requirements for the first 5 years would cost $60 million for OS III and $100 million for OS II. This would require a 20 percent increase in water-user charges—an increase in average annual family water cost from $25 to $30. When augmented secondary treatment became inadequate and tertiary treatment became necessary to meet OS II standards, average family water cost would go up to $37.50, probably within five years. Even such small increases in cost were considered politically unfeasible in Philadelphia. Also, the authorization to issue bonds for construction costs had to be sought separately, one plant at a time, and this magnified the political impact of the expenditures on the city's elected officials.

There is also some question as to what benefits Philadelphia would have received from a higher level of water treatment. A cleaner river would not make it any cheaper to produce palatable and safe drinking water for Philadelphians. There would be a small improvement in aesthetic values and some improvement in recreation possibilities in and near the city. But it is unlikely that an estimate of benefits could have been made which would come anywhere near equalling the cost of $60 to $100 million to Philadelphians.

There was also the problem of the alternative goods (more schools, lower taxes) which would have been foregone by citizens of Philadelphia if the higher cost standards of OS II had been chosen. Voters in the city had turned down three school bond issues between 1960 and 1965, and the school system generally was in need of extensive renovation and upgrading. To judge the true opportunity costs of any of the alternatives would have required information on methods of authorizing and financing alternative programs, on the requirements of the bond market, and on political considerations of the acceptability of various city objectives. Given such data, it is not clear that the resulting calculations would have been accurate enough to be worth doing. But making any judgment (or doing nothing) *assumes* such a calculation. A complicating factor is that the cost of construction of sewage treatment plants had been going up steadily at 5-6 percent per year, so that delay was not costless.

A strictly political question arises as to whether federal assistance would have been more likely if there had been an open fight between the city and DECS over standards, or if there was no fight. Given an uncertain timetable for federal appropriations, the city had to decide which timing of its decision would produce the best chance of being considered for a large share of federal aid. All these decision factors are repeated, of course, for every political unit in the compact, each of which had to vote on the various alternatives open.

312. Outcome

Essentially the same data as shown here were submitted to the three DECS advisory committees and to all subcommittee members. Through a long process involving numerous meetings and communications the chairmen extracted the view-points and expressions from their members and arrived at a consensus for their committees. At their eleventh meeting on March 28th, 1966, the advisory committees arrived at OS III as their compromise recommendation for the development of a water pollution control plan for the estuary.

Of the four control programs discussed (cost minimization, uniform treatment, etc.) a modification of the zoned system was adopted which did *not* contain an effluent fee but rather a waste-allocation formula. The effluent fee provision was dropped apparently on the philosophical basis that it sanctioned contamination of the environment and put the government in the business of licensing polluters and then cleaning up after them.

On deciding that the present oxygen consumption by pollutants (COD) in the waterway must be reduced, the Delaware River Basin Commission decided in 1967 to divide the estuary into four zones, each having a share of the waste-accepting capacity. Each zone's capacity was divided among its dischargers. Within each zone, all wastes were to receive a minimum of secondary treatment (removal of practically all suspended solids and reduction of oxygen-consuming pollutants by at least 85 percent). Except for storm water bypass, discharges containing human wastes or disease producing organisms must first be disinfected, thus protecting river recreation users and shellfish.

In June, 1968, 81 public and industrial dischargers with 92 plants were assigned a maximum permissible oxygen-consuming waste discharge allocation by the basin commission. The allocation was made by the Executive Director after a notice and hearing. A reserve for new dischargers was maintained, and the capacity can be reallocated whenever the reserve approaches depletion or the circumstances render the existing allocation inequitable. Provision was made for progress reports, inspection, surveillance, and non-compliance hearings and citations.

Given that a waste discharge permit system was adopted, it is curious that no consideration was given to making waste discharge permits negotiable. One might consider how such a system might work, what kind of governmental regulation of the discharge permit market might be required, and what existing problems not faced by the plan as adopted might be solved by it.

The commission also adopted the provisions that stream standards could vary within each zone, with higher oxygen requirements for trout waters during periods when anadromous fish make their runs to spawn. Other requirements were set on water temperatures, alkalinity-acidity balance, odor, detergents, fluorides, radioactivity, and turbidity.

Unfortunately the efforts of the commission did not clean up the Delaware—at least as of the summer of 1972. In the 70 miles between Trenton, New Jersey and Wilmington, Delaware, a span that includes Philadelphia, Bristol, Camden, Chester, and Marcus Hook, the river is described by the Federal Water Quality Administration as one of the twenty most polluted in America. At Torresdale, where the city of Philadelphia has an intake of the water which its 2 million persons drink, the river has a coliform count of 4,100. Traces of mercury and lead have been showing up regularly in water samples, and the water itself is a dull brown. Further downstream at the confluence of the Delaware and Schuylkill rivers the DO level is 1.4 ml/liter—far too low for any fish to survive, and the fecal bacteria count is 13,000 per fluid ounce. In the tidal estuary area from Trenton south to Delaware Bay, the DO level is so low that decomposition has become anaerobic, and methane and hydrogen gas bubbles from decomposing matter on the river bottom rise to the surface. The Delaware River Basin Commission attributes much of the problem to the more than 100 factories and refineries along the river above Trenton which dump oil, phenolic compounds,[22] hydrocarbons, and other matter into the water, some through buried discharge pipes which release noxious wastes at night when detection is difficult. The commission remains hopeful that by 1973 or 1974 better technology and stepped up policing will allow its initial quality standards to be achieved,[23] and the Delaware to regain some of its former glory.

313. The Russell-Spofford Model [24]

An operational partial-equilibrium model that is several steps advanced from that used in the DECS has been developed by C. S. Russell and W. O. Spofford. The details of the model have not yet been published. However, its essential components are a linear programming inter-industry model and an environmental diffusion model. The linear programming model relates inputs and outputs of the various production processes and consumption activities within a region, the wastes generated in the production of each product, the cost of transforming these wastes from one form to another (for example gaseous to liquid in the scrubbing of stack gases), the costs of transporting wastes from one place to another, and the cost of final discharge related activities, such as landfill operations.

This model assumes neither that the form nor quantity of the wastes generated are fixed, nor that production processes are predetermined. Thus the inter-industry model permits choices among production processes, raw material input mixes, by-product production, recycling of wastes, and in-plant process changes, all of which can reduce the total quantity of wastes produced. The environmental diffusion model relates the amounts and types of wastes discharged into the atmosphere and water to the amounts and types of wastes that are already present in the various receptors —man, animals, plants, and inanimate objects.

The Russell-Spofford model permits analysis and evaluation of numerous strategies for environmental protection, including effluent charges, discharge standards, and incentives to induce private and public investment in waste control equipment. The analysis can consider all the necessary independencies among liquid, solid, and gaseous wastes in a region. The model has not as yet been applied to an actual region, but will probably be first applied to the Delaware Estuary area. The model will prove tremendously useful if some empirical flesh can be put onto its bare mathematical bones, because it can be the vehicle for demonstrating *with numbers* the points that noneconomists have difficulty grasping from abstract economic reasoning alone.

314. A Managerial Perspective—The Case of the Offending Effluent

To get a feeling for the businessman's approach to the economics, law, and headaches of pollution abatement, consider the case of the Great Marvel Company in Jamestown. The plant manager at Great Marvel got to know officials at the state Department of Health a good deal better than he ever cared to, since the plant was cited repeatedly for violating standards on waste discharges into a stream. The case—which happened to an actual company, public apology and all—was originally prepared by Mr. Ram Charan of the Harvard Business School. The company name and location have, of course, been disguised.

THE CASE OF THE OFFENDING EFFLUENT

"Mr. Palmer, there are two inspectors here from the state Department of Health. They want to see you immediately."

That announcement from his secretary was the first inkling for Edward Palmer that Great Marvel Company's plant in Jamestown had a pollution problem. . . .

Palmer, the plant manager, had been meeting with key subordinates when the secretary interrupted.

Palmer adjourned the meeting. As he walked down the hall to his office, he wondered what they might want of him. He shook hands with the two men and escorted them into his office. The older of the two, who introduced himself as Alan Taylor, began:

"Mr. Palmer, we're inspectors from the Division of Water Quality Control, and we're here to warn you that your plant is operating in violation of water quality standards. You'll have to take corrective action to keep your operating permit. The fluoride concentration of your effluent into Woodbridge Run exceeds the limits."

"On what basis do you say that?"

"We've been taking samples for several weeks," Taylor replied.

Palmer was indignant. "You should have told me you were looking around here," he said.

"You'll get the exact figures in writing from the department in a short time," Taylor said, ignoring Palmer's remark. "The department expects you to comply with regulations as soon as possible."

After the inspectors had left, Palmer telephoned Daniel Jones, the manager of Plant Services at the company's headquarters in New York City, to apprise him of the events and ask his advice. Jones was responsible for providing staff services for environmental control to Great Marvel's plants.

"Those bloodhounds," Palmer said, "have been taking grab samples without letting me know. They treated me like a criminal."

Jones commiserated with him and asked him to arrange a meeting with state Department of Health officials after he had received the written charges from the agency. Palmer agreed.

Three weeks later the promised letter arrived. Dated December 15, 1969, it stated that the fluoride concentration in the waste discharge of the Jamestown plant averaged 3.8 milligrams per liter (m/l). That exceeded the prescribed limit of 1.0 m/l.

The samples indicated that the plant met the department's standards for concentrations of other constituents of the discharge, such as chromium, zinc, and nickel (which the letter itemized). The letter concluded:

"You are hereby directed to maintain a daily operation report to be submitted weekly to this office which will show the volume of discharge and chemical concentrations thereof. The specific chemical analysis must be made daily for pH, total iron, total solids, fluoride, hexavalent chromium, and copper. These analyses should be made on a daily composite sample. An analysis for nickel and zinc should also be made once per week. . . .

"Furthermore, you are directed to formulate a plan for permanent correction of conditions causing violation of state law, and submit it to the Chief, Division of Water Quality Control, Department of Health, no later than September 30, 1970. . . .

Palmer noted that September 30 was the day the plant's current operating permit expired. . . .

Early the following February, Palmer, Jones, and the plant engineer met with government officials at the Department of Health. Among those present were Taylor, the inspector, and Gordon Parry, Engineer for the Bureau of Sanitary Engineering, who had charge of the region that included Jamestown. . . .

"There's been another violation," Taylor said. "We got a complaint last week from the Bureau of Fisheries and Wildlife that several hundred fish had died in Woodbridge Run. We took tests, and they showed that potassium permanganate dumping caused it."

"I'm sorry," Parry said, "but we'll have to fine you in proportion to the number of dead fish. And we think you should apologize publicly through a press release to the newspapers, radio and TV."

"Apologize publicly!"Palmer burst out. "Do we have to?" . . .

Parry replied, "Unfavorable publicity may bring considerable pressure on your company, what with the present widespread concern over pollution. We have to discourage other companies from violating regulations, and this is one way of doing it."

Taylor broke in. "You don't deny that potassium permanganate was dumped from the plant? The water was completely pink."

Jones looked at Palmer, who in turn looked at the plant engineer, who just shrugged his shoulders. . . .

Parry told the Great Marvel representatives that the weekly reports had been quite satisfactory. He said he hoped to hear from them soon about their plans for a permanent solution to the problem. As the meeting broke up, he urged them to keep close watch on the effluent going into Woodbridge Run.

Later, at the plant, Jones and Palmer learned that occasionally potassium permanganate was accidentally dumped into the stream. Palmer told Jones that he had been unaware of a prescribed limit on concentration of the chemical in effluent.

The company paid the fine and apologized publicly. Palmer fervently hoped that the affair of the dead fish was the last such incident until the company had taken permanent abatement measures. . . .

But it was not. Before continuing, however, we should sketch in information about Great Marvel, the town of Jamestown, and the state's pollution control efforts, for it is essential in the understanding of this case to know something about the environment in which the company operates.

Great Marvel: The company is a diversified, decentralized multi-million-dollar enterprise growing at a compound annual rate of 20% over the last decade. An intermediate manufacturer and supplier of components to the auto, aerospace, and electronics industries, it has 16 plants in the United States and subsidiaries in several countries.

The company is divided into three major product groups and supporting staff groups, including Operating Services. Each product group is subdivided into various product divisions.

The Jamestown plant is a key element in the Metal Products Division of the Manufacturing Group. The manager is responsible for its efficient and profitable operation. However, he must secure approval for every significant plant capital investment not only from his divisional vice president but also from the group vice president in charge of Operating Services.

The Operating Services Group, headed by Robert Reller, provides engineering services to the 16 domestic plants. Its task is to identify engineering problems common to most of the product divisions and to plan appropriate action for them. It contracts out most of the engineering work to consultants. The division's responsibilities include dealings with all governmental agencies. This arrangement is designed to permit the plant managers to concentrate on running the plants efficiently.

Plant Services, which Jones manages, is a section of the Operating Services Group. It provides engineering and other services to the plants, with particular attention to environmental control. All of Great Marvel's domestic plants, spread across several states, face pollution problems similar to those encountered by the Jamestown plant. Jones's job is to rectify these situations.

Town of Jamestown: The municipality, population 7,000, is governed by an elected mayor and a council, which includes the municipal engineer. The Municipal Sewer Authority, under the council's jurisdiction, can prescribe what effluents are acceptable to its sewage treatment plant, and at what cost.

Of the 10 manufacturers operating in Jamestown, 2 are heavy industrial. The smaller of these, Doone Corporation, a defense contractor, also discharges its effluent into Woodbridge Run. It contains a high concentration of phosphate and chromium.

Great Marvel's plant was built in 1951 to manufacture defense products. It ships all of its production to customers outside the state. The facility's revenues have grown at 10% a year to about $20 million in 1970; its net income has averaged 5%. During the last 12 months, however, its activity has slackened as defense needs declined. But with 1,400 workers, the plant is still by far the largest employer in Jamestown.

The components it makes undergo precision forging, which requires cleansing of the component after each forging stage. And chemical cleansing treatment—the most common method—produces rinse water that may contain substances harmful to aquatic, plant, and human life.

Recycling of this rinse water is not technically desirable because dissolved nitrates cannot be precipitated, and recycling only increases the degree of nitrate concentration. Other methods of cleansing forgings are either prohibitively expensive or not conducive to producing forgings of required precision. So, Great Marvel uses chemical treatment.

The rinse water is discharged into Woodbridge Run, which empties into the Johnson River about two miles from the plant. Both streams are intrastate, and their water quality control is under the jurisdiction of the state Department of Health.

· · ·

In 1960 the Department began to require every company to obtain a permit in order to discharge waste into a state stream. The permit, renewable annually, enabled the department to collect data about the composition, volume, and pattern of effluent discharge by each permit holder.

The permit initially stipulated that a discharge must not lower the quality of the receiving stream below the U.S. Public Health Service's drinking water standards. This clearly set upper limits on concentrations of such substances as chromium, lead, and fluoride.

•   •   •

In April the group from Great Marvel again met with Gordon Parry. It soon became apparent that the analyses submitted to his office over the last two months, in accordance with instructions, had never been reviewed.

Palmer tried to conceal his annoyance over the waste of time and effort, for he wanted to stay on Parry's good side. Instead, he tried to impress on the regional engineer how diligently the company was working to improve its facilities.

"We're investigating the possibility of expanding the Jamestown operations to include anodizing and nickel cadmium, which are subcontracted out at the moment," Palmer explained. "Then we could treat the total effluent on an integrated basis, because the economies of scale involved in treating extra waste from anodizing and nickel cadmium operations would make it worthwhile."

"That sounds reasonable," said Parry. "But I have another suggestion you might want to look into. Only your fluoride and chromium wastes are above the prescribed limits. Why not separate these from the effluent and discharge them into the Jamestown sanitary system? Then you'd completely avoid the problems associated with discharging into streams.

"By 1973 the town plans to have an expansion that can treat up to 3 million more gallons of effluent a day. All we're waiting for is financing, which is a bit tricky. I won't go into details, but the municipality is hoping for a federal grant of up to 30% of the cost. However, the project will cost approximately $4 million, and we feel we should wait for the present high interest rates to decrease before we try to raise the balance."

"That would be a pretty good solution to our problem," said Jones. "But do you think they could deal with our 450,000 gallons of effluent a day?" Parry assured him that the treatment plant could.

Since 1967 the Great Marvel facility had been discharging waste from its vapor blast area into the sanitary sewer without permission. Palmer and his staff had received no complaints about it. Palmer and Jones agreed that their relations with the chairman of the Sewer Authority were amicable, and they knew that the plant's importance to the community carried a great deal of weight. So they were optimistic that the chairman would accept Parry's suggested solution.

Two weeks later Palmer and his plant engineer had lunch with the Sewer Authority chairman. He appeared receptive to the idea and promised to take it up at an early meeting of the municipal agency. Shortly thereafter, Jones informed Palmer that the company had decided to hire a pollution consultant to prepare a preliminary plan of action.

Changing of the guard: Palmer, to show Parry that progress was being made, telephoned his office in May to arrange another appointment, only to learn that the engineer had just retired. Clarification of Great Marvel's situation in Jamestown was put off for several months until Parry's successor, Harold Smith, had familiarized himself with the job.

Finally, in September, Palmer, Jones, and the plant engineer met with Smith and others from the Department of Health. Smith appeared to be less amenable than Parry; he reminded them that nine months had passed since the department had instructed Great Marvel to come up with a plan for permanent abatement of the pollution.

Jones and Palmer assured Smith that the company was not dragging its feet. Jones said that Great Marvel's consultants had completed their feasibility studies and should have a report ready for corporate headquarters within six weeks.

Palmer said that, because of the illness of the chairman of the Jamestown Sewer Authority, negotiations with that agency were at a standstill. Jones added that in the meantime the plant was bending every effort to prevent illegal discharges into Woodbridge Run, although he acknowledged that it was happening occasionally.

"Well," said Smith, "we don't want to cause you to shut down your plant, so I'm going to recommend to my superiors that your permit be renewed for another year. But remember, that's discretionary, and the permit can be withdrawn any time if it's decided that the company isn't serious about clearing up the problem."

On returning to his office in New York, Jones telephoned the president of the consulting firm and instructed him to have a proposal ready within two weeks.

Late in November Jones's office submitted the plan to the Department of Health. . . .

The plan called for building an integrated disposal system. The investment would be $1 million, but the operating expense would be insignificant, provided that the current volume of output remained constant or increased and the investment was depreciated over 20 years. The system could be in operation within 12 months.

No response to the plan was forthcoming, and Great Marvel continued to operate under its permit. But on December 20 Palmer received a letter from the Department of Health informing him that the pH value of discharge was outside the permissible range of no less than 6.0 and no more than 9.0. The first sample collected on December 4 was 4.1, the second 10.0. The letter concluded:

"You are requested to submit to this office, on or before February 5, 1971, a schedule for correction of the above violation. The schedule must show the dates on which each step will be completed. Failure to submit the schedule as requested will result in Industrial Wastes Permit No. 83-B-54 being subject to suspension.

"This is also to remind you that even though an engineering report for the construction of additional treatment facilities has been submitted to this office, it will still be necessary for you to provide treatment that will meet the conditions of your existing permit."

Palmer passed the news on to Jones, who then instructed him to remedy the pH conditions. This investment amounted to $1,000, with an additional operating expenditure of $100 per month.

•   •   •

Fit to drink: Shortly after Christmas Jones showed up in Palmer's office. "Ed," he said, "I've got some bad news. New York has turned down the million dollar investment."

"Oh, no!" Palmer said. "That's a tough break. Well, I can guess why—because our government business is off, management won't buy that assumption of a 20-year life for the plant."

"Yeah, that's the trouble. They think an eight-year life is more reasonable. But the added depreciation would mean an increase of $75,000 a year in operating expense, which isn't economically feasible."

"Have you got something else in mind?" asked Palmer.

"We're working on a more realistic plan, but it'll take some time. By the way, I've got something else to show you, which isn't very good news either."

"Thanks, I really need something else," Palmer said with a slight smile.

"I just got a letter from the Department of Health laying out some new standards for the plant," Jones said. "They seem to be changing things all the time; even the pH value is different from what it was in that letter you got from them before Christmas."

Jones took a piece of paper from his coat pocket. "In many respects the standards aren't as stiff as they were a year ago, but they still are tighter than for drinking water. Just for kicks I made a comparison. Look at these figures here."

He showed Palmer a table of three sets of criteria for determining water quality (reproduced in *Exhibit*).

Exhibit. Permissible standards for the Jamestown plant's waste discharge, compared with drinking water standards

Type of concentration	U.S. Public Health Service drinking water standards, 1962	State Department of Health standards, December 1969	State Department of Health standards, December 1970
pH	5.0-9.0	6.5-9.0	6.5-8.5
Cyanide	0.2	0	0
Chromium (hexavalent)	0.05	0.02	.025
Nickel		0	0.12
Zinc	5.0	0	0.06
Fluoride	0.8-1.5	1.0	1.75
Nitrates	45.0		
Iron	0.3	1.5	
Dissolved solids*	500	500	

*Parts per million [All figures in milligrams per liter, except for dissolved solids]

"I see what you mean," Palmer said. "Chromium allowed in '62 in drinking water, .05, down to .02, then up to .025."

"You know, even if we made that million-dollar improvement, we wouldn't meet the new standards."

"You're right. But even if we did, the state would change the ground rules with the next grab samples. Look," Palmer said, pointing to the sheet of paper, "the new standards have no entry for iron or dissolved solids. You can bet on it, the reason for no iron is that there wasn't any in the last sample they took in Woodbridge Run."

"No doubt," said Jones. "At any rate, I think we're in for some more trouble with our effluent before we get a new plan submitted."

He was prophetic. In the next few months Great Marvel was fined a total of $33,000 for various violations of the latest permissible limits.

In June 1971 the promised alternate plan, carrying preliminary management approval, was submitted to the Department of Health. It called for an initial investment of $300,000, with an additional annual operating expense of $60,000 after allowance for state tax incentives.

An attractive feature was the speed of implementation; the improvement could be operative within six months of state approval. A disadvantage was the anticipated composition of the discharge which would still not meet the conditions set down in the letter that Jones had received the previous December.

Robert Reller, the vice president who headed Operating Services, at this time was wrestling with the task of drawing up a policy memorandum for his superiors dealing with environmental problems. To stimulate his thinking, he called Jones in one day in September to bring himself up to date on the Jamestown situation. . . .

"In Jamestown we still have a couple of options open to us," said Jones. "Let me lay them out for you and get your reactions. First, there's the revised plan we sent to the Department of Health in April. Though it doesn't meet their regulations, I haven't written it off altogether. There's a chance that an appeal to the governor might get it accepted. Then there's the possibility of putting the effluent into the town's sanitary sewer."

"What proportion of the population works in our plant?"

"About 20%, from in and around Jamestown."

"We could put some pressure on the town to accept that, I imagine," Reller reflected. "How much would it cost?"

"A $300,000 initial investment, and the total expense, including the sewer charges, would increase by $50,000 a year," Jones said. "The drawback to this is that if anything goes wrong at the sewage disposal plant, we're usually seen as the culprits. Just last month the Sewer Authority fined us $12,000 for damage caused by input of a chemical."

"That wasn't our fault, was it?" asked Reller.

"It couldn't have been. We don't discharge even 15% of the total amount treated at the disposal plant. All the same, we had to pay up. We'd have to keep our effluent down to a maximum of 10% of the total throughout to avoid this kind of trouble." . . .

"The Sewer Authority could demand at any time that we treat the discharge in a different way, and it can raise sewer charges at will. Maybe, at some time in the future, they wouldn't even be able to handle the waste at all."

Reller and Jones went on to discuss how they could speed up the necessary modifications to the plant's methods of treating its waste. Great Marvel's operating permit in Jamestown will expire on September 30, so time is short. The constant changes in standards, however, surround the decision with considerable uncertainty.

And Reller is left with the task of drawing up his policy memorandum.

• • •

The business-oriented reader might at this point consider a series of questions on how a situation like that facing Great Marvel might be approached. What is the necessary short-term action required? What short of long-term policy decision is required? What leverage does Great Marvel have, and should it attempt to use its leverage? Has Great Marvel been aggressive enough, and should it be aggressive? Should the process of dealing with environmental problems be centralized or decentralized in the company?

Finally, is there any way that the federal share of financial assistance for the sewer project could be "prefinanced" through existing financial institutions? And if not, should there be a way— should this be one of the aims of public policy in the environmental protection area?

REFERENCES

[1]Much of the material in this section is drawn from Allen V. Kneese and Blair T. Bower, *Managing Water Quality: Economics, Technology, and Institutions* (Baltimore: Johns Hopkins Press, 1968), pp. 13-39.

[2]The presence of such pollutants as cyanide salts or certain pathogens will make the water unfit for recreational activities no matter what the DO level might be.

[3]Testimony of Jacques Cousteau before the Subcommittee on Oceans and Atmosphere, United States Senate, October, 1971.

⁴J. E. McKee and H. W. Wolf, *Water Quality Criteria*, 2nd ed., (Sacramento: California State Water Quality Control Board, 1963).

⁵Not all pollutants are detrimental to aquatic life. A modest level of human waste discharge can promote the growth of algae, which in turn enters the food chain to benefit fishing waters. However, extensive algae growth can be toxic to fish life. Similarly, thermal discharges under some conditions may be advantageous to aquatic life, including fish, by increasing water temperatures during winter periods.

⁶An exception is boron. Traces (less than 1 mg/liter) of boron are essential for all plant growth, but concentrations of several mg/liter result in drastic reductions in yield even for crops tolerant to boron.

⁷Virtually all the models depend on computer simulation. See R. A. Deininger, "Water Quality Management: The Planning of Economically Optimal Pollution Control Systems" (Ann Arbor: University Microfilms, 1965); A. S. Goodman and W. E. Dobbins, "Mathematical Model for Water Pollution Control Studies," *Journal of the Sanitary Engineering Division*, Proceedings ASCE (1966), pp. 1-19; and J. C. Liebman and W. R. Lynn, "The Optimal Allocation of Stream Dissolved Oxygen Resources," *Water Resources Research* (1966), pp. 581-592.

⁸The problem is a variant of one found in J. H. Dale's, *Pollution, Property and Prices* (Toronto: University of Toronto Press, 1968), pp. 27-57. See also John Ayer, "Water Quality Control at Lake Tahoe: Dissertation on Grasshopper Soup," *California Law Review*, Vol. 58 (November, 1970), pp. 1273-1330.

⁹An increase in the number of types of damage to be considered, say from two to twenty, and an increase in the number of ways of avoiding damage from three to thirty (as is realistic with water pollution problems) would produce an enormous matrix, but would not complicate the analysis beyond the capability of a large-scale computer.

¹⁰For the mathematically-inclined reader, the modelling procedure used was as follows. In order to mathematically represent the estuary it was divided into 30 sections. For each of these sections one mass balance equation was written for the biochemical oxygen demand system, and a second equation for the dissolved oxygen system. This resulted in two linear differential equations based on the physical, hydrological, and biochemical characteristics. Once all thirty sections were modelled, the result was two systems of thirty simultaneous equations each. Matrix manipulation techniques were utilized to obtain a set of transfer functions from the coefficients of the equations. The transfer relationships detailed the transformation from a waste load input in any section to the stream quality output in any other section: for example, from effluent waste load to stream BOD, from stream BOD to dissolved oxygen, and directly from effluent waste load to dissolved oxygen.

Some difficulty was experienced because of the technical complexity of the program, and the inability of non-technically oriented persons to comprehend many of the technical aspects. The river mathematical model was sufficiently complex that even the industrial community hired a consultant to verify many of the techniques which were being used. Many persons were said to believe that the whole study was merely an academic exercise preliminary to enforcement procedures which would inevitably follow.

[11]The source of much of this material on the Delaware River Estuary Study is a report by the Federal Water Pollution Control Administration, *Delaware Estuary Comprehensive Study: Preliminary Report and Findings*, Philadelphia (1966) pp. 53-81.

[12]What does the number of objective sets suggest about the extent to which the choices made by political administrators are shaped by their technicians? Does the existence of a small but odd number of choices of objective sets exert pressure for agreement on the middle set? Does this suggest that the administrator in an environmental protection situation must intervene in the problem definition stage to make sure that the technical preparations are geared to the broad objectives of the study?

[13]I am indebted to the discussion of the model in Kneese and Bower, *op. cit.*, pp. 158-164. The mathematical formulation of how a particular DO goal can be secured optimally under each control program is given in E. T. Smith and A. R. Morris, "Systems Analysis for Optimal Water Quality Management," *Water Pollution Control Journal*, Vol. 41 (September, 1969), pp. 1635-1643.

[14]Note that none of these solutions utilizes an explicit measurement of damage resulting from water quality degradation. Rather, the DO level is used as a surrogate which may be considered as a perfectly inelastic damage function.

[15]The inequity could be eliminated by having a central agency construct and operate all of the treatment plants needed in accordance with a cost minimization program, and then distribute costs in relation to the total amount of waste produced. This is similar to the system used in the Ruhr area of Germany, and described in Chapter Five.

[16]This could be redistributed among waste dischargers on the basis of equity criteria, devoted to collective measures for improving water quality, or used for general governmental purposes.

[17]Economies of scale can be obtained by having many small waste dischargers combine their wastes for treatment in a regional treatment plant. To a large extent this has already been carried out in the part of the Delaware amenable to consolidation. All of the City of Philadelphia, some surrounding municipalities, and industries along the river

in the area comprising 40 percent of the waste discharges to the estuary are served by the City of Philadelphia's three treatment plants. Other polluters such as the refineries clustered around the Schuylkill may, because of the nature of their wastes, find it difficult to utilize a municipal plant or other regional waste treatment facility.

[18]One benefit that was not quantified was the value of "playing it safe" ecologically, given the poorly understood and complicated ecology of the estuary area. Damage to ecosystems might be very costly to undo; certainly it might be more costly than the dollar value of preventing such damage now. There is also a strong possibility that ecological damage might occur that would be undetectable for a long period of time, and that would in its long-run effects be adverse to human health. In either case it is clear that OS I or OS II provided substantially more protection against these hazards than did OS III, even where the only difference in river quality is in the DO level. Given that evidence probably cannot be produced to prove or disprove any of the possibilities of damage, the question remains as to what weight should be given to each possibility in evaluating the benefits of each alternative set.

[19]It is not clear what significance the dollar benefit derived under these conditions has. Is there an existing capacity of fishing boats and fishermen, and if so are they currently idle or currently fully employed in other fishing areas? Might benefits from increased fish harvest in the Delaware Estuary be offset by losses in fishing harvest in other areas? For example, what should be the geographical or conceptual boundaries used in calculating the costs and benefits of the various objective sets?

[20]For a much more elaborate cost-benefit analysis of the recreational value of an improvement in water quality in the Delaware Estuary which utilizes a multivariate regression analysis of data from 1352 households, see Paul Davidson, F. Gerard Adams, and Joseph Seneca, "The Social Value of Water Recreational Facilities Resulting from an Improvement In Water Quality: The Delaware Estuary," in Kneese and Bower, *Water Research* (Baltimore: Johns Hopkins Press, 1966), pp. 187-211. Their approach is to use Public Health Service estimates to produce a total cost curve, to use data from the Michigan Survey Research Center to estimate empirical relationships between activity days, socioeconomic characteristics of water recreation users, and measures of available facilities. Using population projections it is then possible to estimate the demand for fishing, swimming, and boating activity for the Delaware Estuary for each year through 1990. The estimates are prepared in two versions, with and without improvement in the water quality of the estuary.

[21]What sort of boundaries does one use when determining the value of parks and picnic areas? It is interesting to identify other kinds of

boundaries implicit in the DECS study. These are implied in decisions from ignoring certain pollution effects to deciding which dischargers are considered to have inputs to the system.

[22]Phenolic substances are produced in the distillation of petroleum and coal products. In concentrations of a few parts per million they do not exert much oxygen demand nor are they very toxic to fish. But even in the smallest concentrations they pose a serious problem to the preparation of drinking water. When water containing minute amounts of phenols is chlorinated to kill bacteria, extremely evil tasting chlorophenols are formed. Also, very small concentrations of phenols in a watercourse can impart an unpleasant carbolic taste to fish which destroys their commercial value as effectively as if the fish were killed. The United States Public Health Service recommended drinking water standards provide for an infinitesimal 0.001 mg./liter limit for phenols. Immediately above Trenton, water samples show a concentration of about nine hundred times this amount.

[23]Part of the policing is from a Gulfstream aircraft operated by the Army Corps of Engineers. The aircraft carries a thermal scanner, ultraviolet and infrared cameras, and polarizing filters to detect sources of pollution. The Corps hopes that the thought of a plane overhead capable of spotting emissions even in the dead of night or through cloud cover will be sufficient to deter most polluters.

[24]This summary of the Russell-Spofford model is taken from Allen V. Kneese, "Environmental Pollution: Economics and Policy," *American Economic Review* (May, 1971), pp. 159-160.

4

Chemical

Pollution

"Hey, Alchemist . . . do you have a magic potion
   that'll kill insects?"
"Sure do, Gort! This one's odorless, and quite
   powerful! . . . It's guaranteed to kill mosquitoes,
   lice, and other disease-carrying pests. I
   call it DDT!"
"How much does it cost?"
"One pebble."
"That sounds too good to be true! What if I try
   your DDT and don't like it? . . . I see you sell
   the antidote. How much do you charge to
   neutralize the effects of DDT?"
"Seventy-eight trillion jillion dollars."

Michael Kelly, in GORT (March, 1971)

400. The Nature of Chemical Pollutants

There are a large number of chemical pollutants currently in use which have known harmful side effects to people and
the environment. Only the major chemical pollutants are discussed
in this chapter.

Some of the harmful effects of chemical and related poisons have
been known for some time; others are still being discovered. In
1967, a study by L. E. Cronin indicated that a concentration of
only .05 ppm of DDT would kill 90 percent of the oysters which
it contacted, and that those oysters which survived may concentrate the toxicant and live to be eaten by higher trophic organisms
that eventually acquire a lethal dose. Cronin found that in a
period of two days, oysters selectively took up 96 percent of the

available DDT and that a large portion of this concentration was retained for substantial periods.[1] More recently, investigators at Harvard have found that chemical pollutants seriously cripple the capacity of marine bacteria to find food, and so may threaten the essential role of bacteria in the ocean food chain. Bacteria have the ability to sense chemicals, and use this sensing ability to find food. The scientists found that many of the bacteria they tested could not detect food when small, non-lethal amounts of organic chemicals such as alcohols were added to the sea-water. There is no reason to believe that the cleansing action of bacteria in fresh water would not also be affected by similar pollutants.[2]

Many chemical pollutants are thought to be carcinogenic, or cancer producing, when human beings are exposed to them. A cancer is a group of living cells, which for reasons not fully understood begins to grow and reproduce without the usual controls. It is strongly suspected that many environmental factors are predisposing causes of cancer. This includes exposure to chemical irritations resulting from cigarette smoking and, to a degree, air pollution.

The concept of tolerance, or "safe" levels of contamination of the environment or of foodstuffs rests on the threshold hypothesis. The threshold hypothesis holds that for all chemicals there is some dosage, greater than zero, at which there will be no harmful effects on exposed individuals. This concept of the tolerance dose was formerly used in the development of radiation protection standards. The accumulation of evidence now suggests that there is no radiation dose that is safe for all exposed individuals and which does not lead to an increase in genetic mutations. The threshold hypothesis has thus been discarded in favor of an alternative hypothesis called the no-threshold hypothesis which holds that *any* radiation dose is accompanied by an increased risk of deleterious biological effects—the magnitude of the risk increasing with the dose. The establishment of standards for radiation protection involves balancing the risk inherent in a particular level of exposure against the benefits to be derived.

The no-threshold hypothesis of dose-effect relationships has also been advanced for chemical pesticidal residues. There is speculation that continued ingestion of small quantities of pesticides could result in serious, but insidious somatic effects upon the more sensitive individuals within a population. Such effects might not become apparent until widespread irreversible damage had been done.

Only in the case of carcinogens is the no-threshold hypothesis currently accepted in United States public policy. The Delaney clause of the 1958 amendment to the federal Food, Drug and Cosmetic Act provides that no carcinogenic substance be permitted in food in any quantity. This clause is extremely difficult to administer, since widely encountered ingredients of food such as common table salt in sufficient quantities are known to be carcinogenic.

The synergistic effects of chemical contaminants are also startling. For example, in mixing the organic phosphorous compounds malathion with triorthecresyl phosphate (TOCP), the combined toxicity of the two is from 88 to 134 times greater than their effect administered singly. Given the existing research effort to achieve more effective pesticides by creating synergisms, a danger of accidental or unintended side effects is always present. The risk to be assessed is not that of a few thousand deaths or obvious cases of genetic mutation (already observed in newborn infants with every fiftieth new basic pesticidal chemical or every ten-thousandth new commercial formulation marketed), but that in the long run we shall all be dead of synergistic poisoning.

MAJOR ENVIRONMENTAL CHEMICAL POLLUTANTS

401. DDT

The chemical which has probably produced the greatest amount of environmental pollution is DDT. DDT was not used widely as a chemical pesticide until World War II. Its reputation as a miracle insecticide effective in controlling tropical diseases such as malaria, typhus, and yellow fever was quickly established. In 1948, the Swiss chemist who discovered its broad insecticidal properties was awarded the Nobel Prize. Spurred by predictions that all major insect pests would be eradicated by this extraordinary chemical, production of DDT in the United States reached a peak of 188 million pounds in 1962-1963. Since then the domestic use of DDT has slowly decreased, largely because of the appearance of resistant strains of insects (which now number 97 species in the United States alone). Most of the DDT produced in the world currently is used for malarial control in underdeveloped nations, but nonmalarial use is increasing rapidly as agriculture in developing countries adopts chemical control techniques. Present high use field crops include tobacco, cotton, peanuts, corn, wheat, hay, apples, potatoes, and most of the berry family.

DDT's unique biological and chemical properties combine to create its extraordinary pollution potential. It is a persistent chemical, having a half-life of up to 15 years. It has extraordinary mobility and is transported by air and water from its original place of application throughout the world's ecosystems. For example, pesticides applied on the southern plains of Texas have been carried through the air by dust particles and traced as far as Cincinnati, Ohio only 24 hours after spraying. It is estimated that one billion pounds of DDT are currently circulating through the world's water and air supply. Being fat soluble, DDT is concentrated in living systems and passed along through food chains. A bird at the top of the food chain may accumulate quantities of the chemical thousands of times in excess of the trace amounts found in the lower organisms on which it feeds. Traces of the chemical have been found in penguins in the Antarctic, in polar bears in the Arctic, in tuna in the mid-Pacific, and in bottom-fish in the south Atlantic. The quantities of DDT stored in human beings—an estimated one-fifth of a gram per person—are far in excess of the amount tolerated in the food supply. Nursing mothers, who must make extraordinary demands upon their body fats, transmit to their children doses of DDT three times the acceptable daily intake set by the World Health Organization.

Supporters of DDT point out that it has contributed to the control of an impressive list of diseases including malaria, filariasis, dengue, yellow fever, virus encephalitis, louseborn typhus, cholera, scabies, and Rocky Mountain spotted fever. In the United States, the incidence of the insectborn diseases of malaria, Rocky Mountain spotted fever, and encephalitis has declined since the beginning of DDT use. In the most dramatic case, the incidence of malaria decreased from 137,000 cases in 1935 to 61 cases in 1969. One estimate put the annual loss of income in the southern states from malaria alone at $340 million a year in the 1930's.

Supporters also point to benefits of insecticides such as DDT in agriculture: increased yields, improvements in the quality of produce, and changes in the timing and location of production. The use of insecticides since 1945 has been associated with yield increases for cotton of from 41 to 54 percent. Similar but less dramatic effects have been achieved with other crops. Insecticides have provided 90 to 100 percent control of livestock predators, such as the cattle grub and horn fly. Insecticides also have important effects in improving the quality of agricultural products, for example by preventing codling moth damage to apples and pears which in the past has led to deterioration of from 50 to 100 percent of annual crops.[3]

Pesticides are used also to reduce the degree of production instability to which the individual farmer is exposed; they are regarded as a form of insurance against crop damage. By reducing the variance of yields, pesticides avoid the need for hedging against yield uncertainty by diversification of crops, and allow a more efficient allocation of resources.

The adverse effects of DDT are the demonstrated health hazard of DDT to people and animals. For example, biologists are becoming aware of the long-term effects of chlorinated hydrocarbons. Animal studies also give us some clues. In a definitive study supported by the National Cancer Institute, DDT was added to the diet of mice and both positive and negative control groups of mice were compared. The frequency of tumors of the liver, lungs and lymphoid organs was four times greater in mice fed DDT than in those in the negative control group. The carcinogenicity was clearly established because DDT caused cancer of the same kind and at approximately the same frequency as did known cancer-causing agents. In studies done at the University of Miami School of Medicine, human victims of terminal cancer were found to contain more than twice the concentration of DDT residues in their body fat as did victims of accidental death. Accident victims carried 9.7 parts per million in their fat—about average for Americans—while the cancer victims contained 20 to 25 parts per million. Concentrations of DDT and its breakdown products, DDE and DDD, were also significantly higher in the fat of patients who died of softening of the brain, cerebral hemorrhage, hypertension, and cirrhosis of the liver than in groups of patients who died of infectious diseases. The histories of the patients in the study showed that concentrations of DDT in their fat were strongly correlated with home use of pesticides, with heavy users having much higher concentrations than light or moderate users.

There is some evidence that the amount of DDT stored in human tissues has not increased over the past decade or so in the United

States, but has reached a mean concentration of 9 to 11 ppm. This compares with observed (1966) concentrations of 3.8 in Canada, 3.3 in the United Kingdom, 2.2 in Germany, 12.4 in Hungary, 19.2 in Israel, and 27.9 in Delhi, India. The critical health variable is now one of long-term toxicity, since the oldest people who have been exposed to high concentrations of DDT since conception are now just in their mid-twenties. Thus, the long-term effects on humans of DDT toxicity—such as reduced life expectancies, or the increased likelihood of cancer or other diseases, may not be known for several decades.

There are various biological control methods—the introduction and establishment of predators, parasites, and pathogens of either insects or weeds and the breeding of resistant strains of plants and animals—which provide alternatives to the use of chemical insecticides. The use of natural enemies boasts some spectacular successes.

One example is the sterilization of large numbers of insects, a technique used in the elimination of the screw worm from the southeast United States. A number of chemosterilants have been found which, when used in conjunction with synthetic attractants, make possible the sterilization of insects in the field and reduce the need for costly breeding facilities.

Canadian biologists also have used small mammals in the control of certain insects. The shrew has been imported to Newfoundland to combat the sawfly; the shrew loves to eat fly larvae, and will consume up to 800 sawfly cocoons a day. In Newfoundland this has led to the destruction of 75 to 98 percent of all sawfly cocoons.

Through both litigation and public information campaigns, the Environmental Defense Fund and other organizations have attempted to eliminate the use of DDT and related pesticides in the United States. The Environmental Protection Agency in Washington did issue an order in June of 1972 banning virtually all use of DDT in the United States after January 1, 1973. However, 27 manufacturers of DDT products immediately appealed the decision through the courts, and a final decision will likely take several more years.

In spite of the substantial amount of literature on the DDT problem which has appeared recently, many of the economic variables relevant to a discussion of the DDT problem are still unknown.[4] There is thus no basis for judging in dollars the increase (or decrease) in productivity attributable to the continued use of DDT.[5] There is no real means of making a first approximation of the economic damage to marine and recreational resources and the toll upon wildlife exacted by pesticides in general, or DDT in particular. Certainly no strictly economic analysis could assign a dollar value to the drastic alternatives concerning human life and the continuation of the species which must be assigned some probability in any discussion of DDT and its effects. We simply don't know whether the price being paid for increased agricultural productivity from DDT use is the sacrifice of the bald eagle through genetic effects related to its reproduction, or the sacrifice or shortening of large numbers of human lives.

402. Mirex

In 1970, eight years after Rachel Carson documented in her book *Silent Spring*[6] the Agriculture Department's campaign to eradicate fire ants in the South with pesticides, the Department had plans to spray 130 million acres of land in nine southern states with Mirex—a chlorinated hydrocarbon considered four times as toxic as DDT—which environmentalists say is like using nuclear bombs on pickpockets. The plan, which would cost $200 million over 12 years, is designed to eradicate the fire ant. The fire ant causes discomfort to picnickers and agricultural farm workers but little or no crop damage. It can destroy pasture land by building mounds six to eight inches high and two feet across.

So far 14 million acres of pastures and woodlands in the South have been sprayed by airplane with Mirex. About 600,000 acres in the Tampa Bay area of Florida were sprayed even though the state's pollution control director complained that the pesticide has a "devastating" effect on crustaceans that live in the Gulf of Mexico. Two million acres were sprayed by the Department around Savannah, Georgia. When the blue crab population along the Georgia coast subsequently died, there were undocumented accusations that Mirex was at fault.

Mirex, like DDT, is highly persistent, has been found as residue in fish and animals, and is a known carcinogen. Traces of Mirex are now appearing in the flesh of beef cattle grazing on lands sprayed with the pesticide. The previous assault by the Department of Agriculture against the fire ant was from 1958 to 1960 when 2.3 million acres in the South were treated with a pesticide called heptachlor. The project was discontinued when birds and fish died by the thousands, and the chemical began appearing in the milk of cattle.

In March of 1971 the Environmental Protection Agency, which

had recently taken over pesticide control from the Agriculture Department, issued notices of cancellation on Mirex use but refused to order a suspension of its sale pending an appeal of the cancellation order, which could take years. Budgetary authority over the Environmental Protection Agency rests in the hands of the House Agricultural Appropriations Subcommittee, the long-time Chairman of which is Representative Jamie L. Whitten, Democrat from Mississippi. By coincidence, the only firm registered to make Mirex for the fire ant program is the Allied Chemical Company in Aberdeen, Mississippi—the heart of fire ant country.

403. Oil Pollution

During 1972, 192 ships carrying cargoes of oil collided with one another, were wrecked on reefs, or otherwise were seriously damaged or sunk. During 1970, the last year for which figures are available, the United States suffered 1,720 oil spills in its coastal and inland waters. Off the coast of Labrador alone, some 325,000 birds a year become so covered with oil that they either cannot fly, or cannot eat without being poisoned because their food supply is coated with oil.

The Alaska Pipeline. Against this background, it is worth noting that seven oil companies began work in 1972 on a project which, if it leads to oil spillage, will make the well-publicized Santa Barbara and San Francisco oil spills look like the oil patch on a garage floor by comparison.

The Department of the Interior filed in late 1970 a report, required by the Environmental Policy Act, on the environmental hazards of the proposed 789 mile hot-oil pipeline from Prudhoe Bay, Alaska, across the Alaskan wilderness area to the seaport at Valdez. The report indicated that each mile of the new pipeline would contain an amount of oil equal to that spilled at Santa Barbara. Only 12 shutoff stations, 60 miles apart, will be installed along the 789 mile route, so a break in the line would produce an oil spill 120 times as great as that at Santa Barbara.

The pipeline will run through a permafrost landscape which is as much as 80 percent frozen. Insulation of the pipe will have to be perfect, because the heat from the oil could turn much of the ground to mud, causing slippage and leaving the pipe supports weakened. The pipeline will climb over the rugged Brooks Range, ford more than a hundred streams and two major rivers, and cross two major earthquake faults that could snap the pipe as easily as a brand new "earthquake proof" hospital in Los Angeles was destroyed in 1971. The tank farm storage complex at Valdez is located in an area that was totally destroyed during the 1964 Alaska earthquake. (The oil companies intend that oil spillage from any tanks destroyed in an earthquake will be contained within an earthen dike; they assume that an earthquake capable of toppling permanent structures would not sever the dike). Tank-stored oil would then be put into ocean-going tankers at Valdez which, in addition to being extraordinarily active earthquake country, is also one of the stormiest and most dangerous harbors in the world. The tankers themselves will be 15 times the size of those that caused the San Francisco disaster; almost as large, and with nearly the volume capacity of the Empire State Building resting on its side.

The effects of a spill in Alaska are far worse than in temperate zones, where oil degrades after a few years, because in the fragile, cold tundra landscape the oil would not degrade and would remain to pollute for centuries. Further, the Alaska oil is of a particularly toxic type. Also, oil spills in Valdez harbor are complicated by the immediate proximity of vulnerable habitats such as bays and estuaries, and by the physically confining effect of the harbor. A few thousand gallons of spilled oil can have an enormously destructive impact on such a closed environment, whereas the same quantity released far at sea would tend to be dispersed with more nominal ill effects.

One example of oil pollution experienced in a closed harbor was that of the Medway Estuary in England, where in September of 1966, the West German tanker Seestern accidentally discharged 1,700 tons of Nigerian light crude oil into the estuary. During the following 36 hours the sheet of trapped oil was swept back and forth by the tides. Virtually every sea bird in the harbor was found dead or incapacitated; the invertebrate life was devastated; hundreds of millions of molluscs were killed by the oil or by the detergent used to clean up the spill; the local shrimp and Dover sole population disappeared and migratory bird utilization of the area dropped by as much as 95 percent for some species. The overall ecological change was profound and apparently permanent.

One U.S. alternative to the Alaska Pipeline was that of unilaterally declaring the Northwest Passage route through northern

Canada to be "international waters," and determining the feasibility of an ocean transport route between the Alaskan oil fields and the eastern coast of the United States using the Passage. The Humble Oil Company in 1969 converted a large conventional tanker, the Manhattan, into a super ice-breaker. The Manhattan made two voyages, but only with the help of two Canadian ice breakers, and even then she sustained significant hull damage. Humble has not publicly shelved the Manhattan project, although future voyages would be in direct violation of the newly-passed Canadian "Arctic Waters Pollution Prevention Act."

The Canadian concern (other than one of sovereignty) is that there is no presently known technique for controlling or dispelling a large Arctic oil spill, since it would seep beneath the ice and congeal. The effects of a large Arctic oil spill might well be catastrophic; temperature variations could occur with unknown long-range results, since the weather conditions of the Arctic exert a great influence on those of the entire northern hemisphere. Oxygenation processes could be disrupted and retarded; breathing holes could be blocked, destroying the primary food sources for carnivorous wildlife. Finally, the only known nesting area of many bird species could be fouled and destroyed.

A more promising alternative would have been to move the Alaskan oil via a railroad or pipeline down the more geologically stable Mackenzie River in Canada the 1,000 miles to the present railhead at Great Slave Lake. From there, three 100-car trains per hour could have carried 2 million barrels of oil per day to the terminus of an existing oil pipeline in Edmonton, Alberta. The Mackenzie River route would have cost an estimated $2.5 billion, compared to an estimated $3 to $4 billion for the trans-Alaska route. The environmental impact statement filed by the Department of the Interior does not mention the Mackenzie River alternative at all, apparently because the pipeline or railroad would have to run through Canada, and would have partial or full ownership by Canadians, thus denying U.S. oil companies returns from transporting the oil as well as from refining and marketing it.

There are two other alternatives to the Alaska pipeline which received little attention from the Interior Department. The simplest one was to ease trade barriers against foreign oil imports, and leave the Prudhoe Bay oil safely underground until it was really needed. The second alternative was to consider whether there should be a concerted effort to produce and use even more oil since there is not a single use of oil, from cars to power plants to plastic bottles, which does not produce pollution. The Department of the Interior's proposal to develop Alaskan crude oil was based on the assumption that the United States would quadruple its oil demand by the year 2000; that the United States could not depend upon other sources of oil; and that there was no need to contemplate less extravagant use of oil if the Alaskan reserves were developed. Consider, however, that automobiles account for 40 percent of the oil use in the United States. If every American did not use his car just one day each week, by taking a carpool, bus, train, bike, or walking, that would save enough oil in six years to equal the entire Prudhoe Bay discovery. If U.S. airlines curtailed one quarter of their flights (which is feasible, given that their average passenger load is less than 55 percent) they would save enough oil in eight years to equal the Prudhoe discovery. If the proposed French-English and Russian fleet of SST aircraft are not introduced, the Prudhoe Bay discovery will be "saved" in nine years (the Anglo-French Concorde SST requires about 2½ times the fuel per passenger mile as the Boeing 707). Finally, consider that Americans throw away about 100 billion containers each year, virtually all of which are made with oil as an ingredient, or as energy. The same is true with the throwaway cars that we are used to. If some things were recycled, and if others were required to be made so that they would last longer or could be reused, the U.S. need for oil could be drastically cut.

Offshore Drilling. In recent years a new source of environmental contamination by oil has developed. There are now over 9000 oil wells pumping from submerged areas of the world's continental shelves. There are also 150 mobile oil drilling rigs actively involved in exploring for offshore oil. The extraordinary series of oil pollution incidents which followed the discovery of oil in Cook Inlet, Alaska are indicative of the dangers in offshore production. These incidents now average one per week and their severity ranges from modest to extremely destructive. Tens of thousands of sea birds and water fowl have been killed by oil pollution, and concern is growing for the welfare of the mammal populations of the region, including the beluga whale, seals, sea otters, bears, and other furbearers. It has been reported that an entire oil drilling rig was carried away by ice in Cook Inlet. If one adds to this the problem of breaks in the submerged pipelines serving offshore wells, the magnitude of the potential for environmental contamination becomes clear. Such underwater pipelines have been broken by ice action at a water depth of 250 feet.

Release from Ships. There are a few geographic areas where the principal damage due to oil contamination arises from the dumping of oil by ships at sea. For example, most traffic from Northern Europe, the Mediterranean and Africa to Halifax, the Great Lakes and New England ports converges south of Cape Race on the Grand Banks of Newfoundland. Oil dumped south and east of Newfoundland is carried toward the coast by prevailing winds and currents. The concentration of oil pollution results in thousands of dead and dying murres and other sea birds found along the shores of both sides of Newfoundland's Avalon Peninsula in the fall, winter, and early spring.

404. Radiation and the SST

Although the United States Senate has voted down funds for the American supersonic transport plane (SST), both the Soviet and Anglo-French SSTs are being test flown and are scheduled for commercial service during this decade. One of the principal arguments of those opposed to the SSTs aside from the fact that they use so much oil, which we have already noted, is that nitric oxide from their exhaust might set in motion a series of chemical reactions that could destroy the ozone layer of the stratosphere. SSTs must fly in the lower stratosphere at a height of 12 to 13 miles to be efficient. With less ozone to absorb harmful ultraviolet radiation from the sun, it is generally accepted that enough radiation would reach the earth to greatly increase the incidence of skin cancer.

The stratosphere is isolated from the lower atmosphere, or troposphere, in part because of the presence of ozone in the stratosphere. Ozone is a form of oxygen with three atoms (O_3) which is heated as it absorbs solar ultraviolet rays, making the stratosphere warmer than the air below it. When hot air lies above cold air, the cold cannot rise through the warmer layer, which serves as a lid. Thus, the stratosphere is protected from the turbulence below, except on rare occasions when giant thunderheads break through the barrier. Stratospheric air makes for smooth flying; it is also very thin and offers little aerodynamic resistance. However, to date only a few planes such as the SR-71 reconnaissance aircraft have flown that high—existing jet airliners are most efficient in the troposphere, at altitudes of from 5 to 7 miles.

As ultraviolet light penetrates the stratosphere it is absorbed by ozone, which in the process is split into oxygen gas molecules and individual oxygen atoms. The ozone is replenished by a reverse reaction in which single oxygen atoms merge with the two-atom oxygen in the presence of some other atom or molecule acting as a catalyst. Nitrogen oxide robs ozone of one oxygen atom, converting it into oxygen gas which does not provide a complete ultraviolet shield. In this reaction the nitric oxide becomes nitrogen dioxide, with two oxygen atoms; another reaction quickly converts nitrogen dioxide back to nitric oxide, which is ready to attack another ozone molecule. Through this process, it is feared that SST exhaust would first wipe out virtually all ozone in the lower stratosphere, then diffuse up through the rest of the ozone region, continuing to deplete it even if the nations involved suddenly halted flights in the stratosphere.

Some scientists have argued that the breakdown of the heating effect in the stratosphere which keeps it protected from the turbulence below would effect the world's climate and precipitation levels in unpredictable ways. Dr. Harold Johnson of the University of California at Berkeley predicted in a White House paper that a fleet of 500 SSTs operating an average of seven hours a day (the predicted rate by 1985) could reduce the ozone content of the atmosphere by half within eight months. He estimated that each SST would dump one ton of nitric oxide an hour into the stratosphere. As a result, he predicts: "all animals of the world. . . . would be blinded if they lived out of doors during the daytime." Other scientists predict that the ultraviolet radiation would kill all plants except those under water. Dr. Johnson, who is credited with being the first person (in 1950) to measure the rate at which nitric oxide breaks down ozone, says his calculations hold under a full range of variables of temperature and various concentrations of oxygen atoms and nitrogen oxides. When news of Dr. Johnson's estimates became public Senator Edmund S. Muskie of Maine, at a speech at the United Nations on May 21, 1971, urged that no nation fly its SSTs until their effect on the ozone level could be determined by some international group. "We cannot afford to let one nation decide for all mankind," said Muskie.

405. Nuclear Wastes

Nuclear reactors produce three types of radioactive effluents, called Radwastes, which are of concern as possible sources of very persistent pollution. The first of these is gaseous radioactive waste, produced when some gaseous atoms leak through the metal casings of fuel rods in water-cooled reactors. Accumulation of these radioactive gases poses a radiation hazard when men have to do maintenance work in the reactor structure, so the gases are collected, stored, and then released into the atmosphere by venting through a tall stack at a controlled rate. All such releases are required to be less than Atomic Energy Commission (AEC) standards, and the reactor sites are continuously monitored with sensitive instruments to record radiation doses on the ground. The AEC claims that the dosage from atmospheric venting at the boundary of a large reactor does not exceed 0.005 Röntgens (Rs), where AEC standards state that 0.17 R per year per person is considered a safe dosage.

The second kind of radioactive effluent is liquid Radwastes, which are usually included in water that circulates through the nuclear reactor core. A variety of radioactive substances are produced within the reactor vessel and get circulated in the water. To aid in maintenance, these radioactive contaminants are removed periodically, and are released in a controlled fashion into the discharge water of the plant. Again, the levels of radioactivity are subject to strict control, and the radioactive concentration of different atoms is measured in fish, plants, birds, and marine life such as clams and oysters on a regular basis. The AEC states that the radiation doses delivered to communities near nuclear reactors are very small, considerably under permissible yearly dosages.

The third and most controversial kind of radioactive effluent arises from the disposal of the atomic fuel once it is used up. Because of the tendency of some atoms to accumulate in fuel rods of the reactor and rob neutrons from the chain reaction, it is necessary to remove some of the fuel rods before complete burnup of the uranium fuel. Refueling a reactor occurs about once a year. Each used fuel rod removed from the reactor core is immensely more radioactive than radium. The "hot" rods are first transferred to a deep pit of water where they slowly cool. After some months the rods are transferred to heavily shielded caskets and trucked to a fuel reprocessing plant, where they are sheared into small pieces and chemically processed to separate the nuclear fuel from the highly radioactive gases (including krypton-85, which has a half-life of ten years) which are simply vented into the atmosphere. The AEC stores the remaining, noneconomic wastes in special underground tanks; by mid-1972, some 81 million gallons had been accumulated.

The ultimate disposal of these wastes has not been decided; one possibility is to simply dump them in a deep part of the ocean. The current intention is that the AEC will take its wastes, draw off the liquid and form the residue into ceramic blocks, to be shipped to Lyons, Kansas, and buried several hundred feet down in a salt mine. The central Kansas site was chosen because of its great geologic stability. Each of the ceramic blocks will contain about 200 billion Rs of radioactivity. These capsules will be extremely hot, with outside temperatures of from 200 to 400 degrees centigrade and inside temperatures of up to 1,800 degrees. The heat is generated by the decay of radioactive materials. Because many of the radioactive products are extremely long-lived, the storage site would have to be maintained for thousands of years.

In a report filed with the AEC in December, 1970, Governor Robert Docking of Kansas charged that the AEC had "exhibited remarkably little interest" in studying the effects of radiation and heat on the salt and complained that plans for removing the radioactive substances if something should go wrong "do not exist at all."

Perhaps the most troubling long-term problem in all this is that, in every investigation into the topic of nuclear pollution, whether by atomic wastes or underground tests by the AEC, there are almost insurmountable problems of proof. In every case defenders of the environment must depend for support on AEC scientists who frequently will not commit themselves, and on courts which are reluctant to deal with anything in the sensitive area of national defense.

In the recent *Crowther* v. *Seaborg* case,[7] a suit to enjoin the AEC from proceeding with underground atomic tests in the Rulison fields near Grand Junction, Colorado, the court put the burden of proving environmental damage on the plaintiffs, but denied the plaintiffs the right to obtain information from the AEC that the court found the AEC to be "almost exclusively in possession of." The court pointed out that what was involved was a value-judgment "which is in our opinion reserved for institutions within our government framework, not the courts."

406. Solutions To Major Chemical Pollution

One of the common links distinguishing the various kinds of chemical damage to the environment from the cases of air and water pollution is that it is difficult to envision how the externality problems involved in chemically-caused ecological damage might be handled by economic approaches. In the cases of DDT and other pesticides, oil pollution, the SST, and nuclear wastes, the scope and impact of cost externalities almost defies measurement. If significant externalities do exist, for example with the ozone destruction potential attributed to the SST, the scale of possible damages is so astronomical that a cost-internalization approach becomes meaningless. The most feasible approaches to chemical pollution of the environment probably lie in legal cost internalization, and in legislative approaches such as that presented by the National Environmental Policy Act. A discussion of these approaches is found in Chapters Six and Seven.

An alternative solution would be to develop some kind of distant early-warning system that would identify chemical pollution episodes before they have passed the point of no return. The most practical and economical early warning system would be very distant indeed—a monitoring system of space satellites which could detect oil contamination of the seas sufficient to destroy oceanic plant life, pollution of the atmosphere by carbon dioxide sufficient to bring about alteration of the earth's climate, and changes in the vitality of crop or forest lands. In addition to watching man-made perils, the satellites could monitor rises in subterranean heat indicative of volcanic eruption or follow wobbles in the earth's spin that are believed to precede major earthquakes.

The first in a series of satellites designed to test such roles was launched by the United States in May of 1972 in the Earth Resources Technology Satellite program (ERTS). The three television cameras in the ERTS satellite scan the landscape at three different wave lengths selected to provide maximum information on types of natural vegetation and crops, vegetation vigor (as an index of ground moisture), and pollutants. Scanners sensitive to four parts of the infrared spectrum are used to identify the nature of the surface from which the radiation is coming. The satellite will thus be able to detect how much chlorophyll—the substance that makes plants green—is in ocean water, because its presence is a rough index of how much life exists in any part of the ocean. The infrared scanners can also monitor the global distribution of oil slicks. However, to do these things properly, about one hundred satellites will be required—and if they are Soviet and American, the method of collection by spacecraft must be sufficiently uniform to produce a composite picture.

MAJOR INTERNAL CHEMICAL POLLUTANTS

407. The Extent of Internal Pollution

Although the public is becoming increasingly aware of environmental pollution, few people have yet become concerned with a comparable trend in our internal environment—our bodies. Internal pollution occurs in two ways: from the additives and impurities in much of what we eat, drink, and breathe; and from the multiplicity of drugs consumed daily for other than strictly therapeutic reasons. For many the solution to coping with the stresses of urban industrial society is to take one drug or another—a solution which is encouraged and abetted by the advertising media. Pills are available for every contingency: to induce sleep or to prevent it; to reduce weight or to build up muscle power; to achieve pregnancy or to avoid it; to ease pain; to calm nerves.

The equilibrium of the body is very vulnerable to the action and interactions of drugs which overextend its metabolic machinery over prolonged periods. In most cases, the precise effects of these drugs on body systems over the long term are not known.

In testing drugs such as birth control pills whose use is not intended as medication, and where the drug is taken over a very long period of time, it should also be important to establish the effect of the drug not only on the target organ, but on all the vital organs of the body—the liver, kidneys, brain, etc. It is surprising but true that this aspect of drug safety has not only sometimes been ignored, but on occasion, where adverse side effects have been discovered, the news has been withheld from the public. Senate investigating committees find themselves evaluating conflicting expert opinions and the pharmaceutical manufacturers are obliged to send more and more "Dear Doctor" letters of caution to dispensing physicians, but neither contributes in any way to a solution of the basic problem of lack of information.

The sections which follow discuss a selection of "involuntary" and "voluntary" internal pollutants: mercury, cadmium, cyclamates, birth control pills, and cigarettes. Some solutions appropriate to internal pollutants are suggested.

408. Mercury

Mercury, also known as quicksilver, is a substance that, by its natural chemical action, damages the structure of the body or disturbs its functions. Mercury associates with red blood cells and nervous tissue to reach the brain and central nervous system. It easily passes the placental barrier to concentrate in the fetus. In attacking the brain, mercury causes neurological disorders such as blindness, paralysis, and sometimes insanity. It also attacks vital internal organs such as the liver. Swedish scientists report that they have found chromosome breakage in humans who ingested high levels of methyl mercury in fish. These scientists predict that mercury ingestion could lead to genetic damage to future generations.

Mercury ingestion can also be fatal. Fish and shellfish heavily contaminated with methyl mercury near Minamata, Japan are reported to have caused 111 cases of mercury poisoning between 1953 and 1960. Of this group, 43 persons died. Twenty-three of the surviving women bore 19 babies with congenital defects. The mercury was traced to waste from a large chemical plant on Minamata Bay. In a second episode in Niigata, Japan in 1961, mercury pollution of fish poisoned 30 persons, six of whom died. The concentrations of mercury in fish in Niigata suggested that regular, daily consumption of fish containing 5 to 6 ppm of mercury would be dangerous.

The best known source of mercury contamination occurs in the manufacture of caustic soda, one of the basic raw materials for the chemical industry. Mercury is also used as a catalyst in the manufacture of such plastics as urethane and vinyl chloride. The electrical industry uses a million pounds of mercury each year in long-life batteries, which are eventually thrown out. Many seed dressings, fungicides, and sprays used in agriculture also contain mercury.

Although the effects of mercury have been recognized since 1957, no warnings were given to the American fish consumer until December of 1970, when the Food and Drug Administration recalled one million cans of tuna fish from the market because of suspected high mercury levels. The agency insisted, however, that the product was still safe to eat. At that time Dr. Charles C. Edwards Jr., Commissioner of Food and Drugs, estimated that on the basis of samplings, 23 percent of the 900 million cans of tuna fish packed in the United States during 1970 contained amounts of mercury that were considered excessive. The agency had been operating with a mercury tolerance level of 1 ppm; the highest level of mercury found in the 138 samples they tested was 1.12 ppm, with the average of all samples at .37 ppm.

Consumers have also not been told either by the government or by the industry that a number of other countries have set much stricter levels for mercury contamination in fish. The level of 0.5 ppm set by the Canadians and later accepted but not enforced in the U.S., would condemn about 40 percent of the canned tuna currently sold in the United States. Further, Canadian and other officials are seriously concerned about the long-term dangers involved in the ingestion of even small amounts of mercury, and are considering lowering their permitted contamination levels to 0.2 ppm—a level which would condemn virtually all the tuna and swordfish now sold in the United States.

409. Cadmium

Like mercury, cadmium is a heavy metal that only recently has been found to pose an environmental hazard. Cadmium is a bluish-white metallic element which occurs in small quantities in zinc ores. It is used in electroplating, in the manufacture of fusible alloys, and in the control of atomic fission. Cadmium poisoning affects the bones of humans, reducing their calcium and making them so brittle that, just by coughing, a victim can break his entire rib cage. The connection between cadmium poisoning and a disease the Japanese call *Itai-itai* ("It hurts, it hurts") was discovered by Dr. Noboru Ogina, a Toyama physician, in 1960. For about 50 years, farmers in the Toyama Prefecture, or Province, had complained of a disease that seemed to shrivel their bones, and which was fatal in about 50 percent of cases. The cadmium wastes in question had come from drinking water drawn from the Jintsu River, and discharged from the Kamioka refinery of the Mitsui Mining and Smelting Company, which processed lead and zinc mined in the area. In 1967, 14 victims of the disease brought suit against Mitsui Mining, charging the company with responsibility for their illness. Nine of the plaintiffs died during the two year trial; the survivors were awarded the equivalent of $158,000 as compensation; 1,800 additional suits against Mitsui are still pending.

Since 1968, cadmium metal in concentrations which were considered potentially unsafe by the United States Public Health Service have been found in oysters taken along the Atlantic coast of the United States from Maine to North Carolina. The danger was officially reported to representatives of the shellfish industry, officials from the Atlantic states, and federal officials. However, no consumer warning was issued by either government or industry, and no attempt was made to limit or suspend the sale of the possibly contaminated shellfish. The Public Health Service has reported that the average level of cadmium in oysters is 3.1 ppm, with a range of 0.1 to 7.8 ppm over several thousand samples—only one tenth of the cadmium concentration which led to fatalities in Toyama. The current Japanese safety standard for cadmium in foodstuffs is 0.4 ppm; the U.S. Public Health Service has proposed a limit of 2 ppm for cadmium, mercury, lead, and chromium *combined*. No determination has been made of how many oysters would have to be eaten to cause illness, but the number is apparently quite high.

410. Cranberries and Cyclamates

The 1958 Delaney Amendment to the Pure Food and Drug Act specifically denied the Secretary of Health, Education, and Welfare the discretion to approve the use of carcinogenic food additives. The Delaney Amendment provided that:

[No] additive shall be deemed to be safe if it is found to induce cancer when ingested by man or animal, or if it is found, after tests which are appropriate for the evaluation of the safety of food additives, to induce cancer in man or animal . . .[8]

Shortly after the enactment of the food additives law, the Delaney Amendment gained notoriety because of its application in the infamous "Cranberry Incident." In February of 1959, the Food and Drug Administration (FDA) received a joint petition from Amchem Products and American Cyanamid requesting approval of a tolerance of 1 ppm of aminotriazole, a weedkiller, on apples, pears, and cranberries. Such a tolerance would permit minute residues of the chemical to appear in the cranberries themselves. The FDA refused to approve the petition requesting the tolerance because its Division of Pharmacology had concluded from pathological studies that aminotriazole was a carcinogen. Rats which had been fed the substance for 2 years at 100, 50, and 10 ppm had produced thyroid tumors which appeared at the 68th week of feeding in half the rats which had been fed 100 ppm, and in smaller numbers of rats which had ingested 50 and 10 ppm. A "no effect" level was not established.

Cranberry growers had been so confident that the weedkiller would get clearance that they went ahead and used it during harvest. In 1959, cranberries containing residues of the chemical began turning up in various parts of the country. The FDA began seizing the contaminated cranberries, and Arthur Flemming, then the Secretary of HEW, cited the Delaney Amendment as the authority for this action. While cranberry farmers complained that feeding a rat 100 times the expected human dosage was equivalent to a human eating 2,130 pounds of cranberries per year, the Department continued to go beyond simply warning consumers of the possible effects of aminotriazole, and seized most of the affected crop.

In a more recent application of the Delaney Amendment on October 18, 1969, Secretary Robert Finch of HEW banned all further use of the artificial sweetener cyclamate. Secretary Finch stated that recent experiments conducted on laboratory animals had disclosed the presence of malignant bladder tumors in animals which had been subjected to strong dose levels of cyclamates over long periods. The government's action in banning cyclamate seemed to cause much less furor than that surrounding the application of the Delaney Amendment to the "Cranberry Incident" ten years earlier. One exception was a puzzled statement by the President of Coca Cola, who pointed out that the laboratory concentrations of cyclamates used in experiments were the equivalent of a human drinking 250 cans of Fresca or Tab per day for two years, and expressing amazement that fifteen years of safe human experience with cyclamate were being disregarded. The Department's order was carried out, and general purpose foods and beverages containing cyclamates were withdrawn from sale over a three month period. Cyclamates continue to be used in diet foods purchased under prescription, but the user is informed of the possible risk involved in their consumption.

411. The Pill

Oral contraceptives present society with problems unique in the history of human therapeutics. Never before have so many people taken such potent drugs voluntarily over such a long period of time for an objective other than the control of disease. While the birth control pill does furnish almost completely effective contraception, it shares with virtually every other drug some risk factor, albeit minimal.

A whole series of adverse reactions have been observed in retrospective studies of patients receiving oral contraceptives.[9] The most important of these include nausea, vomiting, breakthrough bleeding, edema, breast changes including enlargement and tenderness, loss of scalp hair, hemorrhagic eruption, migraine, allergic rash, rise in blood pressure, and mental depression.[10] British prospective studies released in 1968 showed that birth control pills sometimes resulted in thromboembolic episodes, or blood clots. The British data from controlled studies indicated a seven to tenfold increase in thromboembolic deaths and diseases among oral contraceptive users as compared with other women.

The potential carcinogenicity of oral contraceptives has neither been confirmed nor ruled out.[11] Much indirect evidence suggests that steroid hormones, particularly estrogen, may be carcinogenic in man. The only data available is from experiments on laboratory animals in which long-term administration of estrogen resulted in cancer in five species. There is strong suspicion, but as yet no proof that the carcinogenicity of estrogen in other species can be transposed directly to man. The principal obstacle to documenting proof is the long latent period between the administration of a known carcinogen and the development of cancer in man. So far, no prospective or retrospective studies have been designed to provide an adequate resolution to this problem.[12]

One further problem exists for the women who neglects to take a contraceptive pill, conceives, but not knowing that she is pregnant returns to regular use of the pill. What she has done, according to

Dr. Allan C. Barnes, chief gynecologist and obstetrician at the Johns Hopkins Hospital, is to expose the fetus, if it is female, to an "optimal hazard"—that the fetus will be masculinized by the norethindrone (Norlutin) content in the pill.

Labeling Requirements. For the first fifteen years of their use in the United States, no consumer-directed warnings of any kind appeared concerning the possible side effects of birth control pills. The reasoning apparently was that users would be needlessly frightened, and that the proper source of information on the effects of drugs was the prescribing physician and not the manufacturer. In 1966, the FDA implemented one of the recommendations of the *Report on Oral Contraceptives* and began including in its packaging a warning to users to "Discontinue medication pending examination if there is sudden partial or complete loss of vision . . ." and "for any patient who has missed two consecutive periods, pregnancy should be ruled out before continuing the contraceptive regimen." After receipt of the 1968 British data on the incidence of blood clots, the FDA changed its regulations to require a "patient insert" for oral contraceptives which was 600 words in length, contained a warning about blood clots, and mentioned specific symptoms that should alert users to discontinue the pill.

Three weeks later, and after "talks with doctors, manufacturers and Planned Parenthood," FDA released a new draft version of a patient insert which contained the following 96 words.

> Oral contraceptives are powerful, effective drugs that should be taken only under the supervision of a physician. As with all effective drugs, they may cause side effects in some cases and should not be taken at all by some. *Rare instances of blood clotting are the most important known complication of the oral contraceptives.* These points were discussed with you when you chose this method of contraception.
> While you are taking this drug, you should have periodic examinations at intervals set by your doctor and should report promptly any change in your state of health.

412. Cigarettes

Scientific investigation into the association of tobacco use with various diseases began as early as 1900, but relatively little research was done until 1939, when the first controlled retro-spective study of smoking and lung cancer was carried out. Similar work was published in 1943, 1947, and 1948; twelve more major studies followed from 1949 until 1955. Virtually all of this research tended to show that cigarette smoking is a cause of lung cancer and other diseases. Finally on January 11, 1964 the Report of the Surgeon General's Advisory Committee on Smoking and Health was released. Its judgment was that "Cigarette smoking is a health hazard of sufficient importance in the United States to warrant appropriate remedial action." [13]

Specifically, the Advisory Committee concluded that cigarette smoking is associated with a 70 percent increase in the age-specific death rates of males, and to a lesser extent with increased death rates of females. The larger the number of cigarettes smoked daily, the higher the death rate. For men who smoke fewer than 10 cigarettes a day, the death rate from all causes is about 40 percent higher than for non-smokers. For those who smoke from 10 to 19 cigarettes a day it is about 70 percent higher; for those who smoke 20 to 39 cigarettes a day it is 90 percent higher, and for those who smoke 40 or more, it is 120 percent higher. Smokers who stop smoking have a death rate about 40 percent higher than non-smokers, as against 70 percent for current cigarette smokers. Compared with non-smokers, the mortality risk of cigarette smokers increases with the number of years of smoking and is higher in those who stopped after age 55 than for those who stopped at an earlier age.

Cigarette smoking was found by the Advisory Council to be causally related to lung cancer in men; the magnitude of the effect of cigarette smoking far outweighed all other factors. In comparison with non-smokers, average male smokers of cigarettes have ap-proximately a 9 to 10-fold risk of developing lung cancer and heavy smokers at least a 20-fold risk. The data for women, although less extensive, point in the same direction.

Cigarette smoking was also found to be the most important cause of chronic bronchitis in the United States, and is associated with an increased risk of dying from pulmonary emphysema, but the relationship has not been established as causal. For the bulk of the population of the United States, the importance of cigarette smoking as a cause of all chronic bronchopulmonary disease is much greater than the importance of atmospheric pollution.

The Tobacco Research Council, a cigarette industry sponsored group, has provided the most elaborate disclaimer of the research findings on the relationship between cigarette smoking and health. The Council argues that it is complex, genetically innate individual differences in susceptibility and nonsusceptibility that affect the imbalance of various organ and tissue functions and that lead to major public health problems such as cancer and cardiovascular disease.[14] In other words, the Council argues that there are predisposing factors in one's heredity that may lead him both to smoke and to have a higher incidence of lung cancer or other diseases.[15] This is offered as an explanation of why people who have never smoked continue to die from the same diseases as do those who do smoke, and people who smoke do not die from the statistically implemented diseases. Thus, daily life practices such as exercise, hours of sleep, use of coffee, tea, tobacco, alcohol, and the common household drugs, exposures to all kinds of materials and substances on the job, psychological characteristics, and the stresses of daily living may play a role in the incidence of disease. The Council report goes on to discuss in detail predisposing factors other than cigarette smoking associated with cancer of the lung, the cardiovascular diseases, chronic pulmonary diseases, and central nervous system disorders.

Economics of The Cigarette Industry. The economic significance of tobacco production and manufacture to the American economy is striking. About 700,000 farmers and manufacturing workers are dependent in whole or in part on the tobacco crop, and their combined incomes from tobacco approximate $1.7 billion per year.[16] Tobacco farms are found in 22 different states, as far north as Massachusetts, as far south as Florida, and as far west as Wisconsin and Missouri. The value of manufactured tobacco products amounts to about $5.1 billion per year. Tobacco is an important industry in several states, notably North Carolina and Virginia, where tobacco manufacturing is the second and third ranking industry respectively. The tobacco industry annually buys 40 million pounds of cellophane, 70 million pounds of aluminum foil, 27 billion printed packages, and nearly 3 billion cartons. Tobacco manufacture also includes paper mills, machinery plants, truckers, wholesalers, vending machine companies, advertising agencies, advertising media, and about 4,500 wholesale firms and 600,000 retailers. The federal government, the states, and local government derive about $4.3 billion in tax revenue each year from cigarettes.

The United States is not only first in world tobacco production, but is also the world's leading tobacco exporter. In 1963, the value of U.S. exports was $686 million, and the value of imports $160 million, which means that more than a half billion dollars of trade surplus and balance of payments account was contributed by the tobacco industry.

Labeling Requirements. Cigarettes, like the Pill, also have required labeling of their potential hazards. One week after the release of the Surgeon General's report in 1964, the Federal Trade Commission issued a notice of proposed rule-making intended to affect the labeling and advertising of cigarettes. On July 27, 1965 the Cigarette Labeling and Advertising Act of 1965 became law.[17] The most significant portion of the law affected labeling of cigarette packages, and required that:

> It shall be unlawful for any person to manufacture, import, or package for sale or distribution within the United States any cigarettes the package of which fails to bear the following statement: "Caution: Cigarette Smoking May Be Hazardous to Your Health."

The logic behind the labeling requirement was that while the individual smoker must be safeguarded in his freedom of choice— the right to smoke or not to smoke—he also had the right to know that smoking may be dangerous. The cigarette industry agreed to this warning, and to the absence of any qualifying adjectives such as "excessive," "continual," or "habitual," in return for a clause in the Act which stated that no other federal authority (including the FTC), state or local authority could require any stronger health statement on packages than that required under the Act.

The cigarette industry managed to win its fight against a proposed FTC rule that would have required the health warning of potential hazards to be included in all advertising of cigarettes. In return, the industry agreed that all cigarette advertising would be subject to an industry advertising code administered by former Governor Robert B. Meyner of New Jersey, who would have the power to prohibit any advertising which he determined did not meet the requirements of the code. As in the case of the labeling of cigarette packages, the Act preempts all federal, state, and local authorities from requiring any health statement in the advertising of cigarettes.

Although the 1965 Act precluded any direct regulation of cigarette advertising, the Federal Communications Commission (FCC) did manage to invoke the fairness doctrine as applicable to cigarette advertising. The FCC ruled in 1969 that cigarette advertisements clearly promoted the use of a particular cigarette as attractive and enjoyable, and that a station which presented such advertisements had the duty of informing its audience of the other side of this controversial issue—that however enjoyable, such smoking may be a health hazard.[18] As a result of this ruling, broadcast stations which carried cigarette commercials were also required to present in prime time, anticigarette messages prepared by the various interested health organizations. As the 1969 expiration date of the 1965 Act approached, the FCC issued a notice of proposed rule-making the effect of which would be to ban all cigarette advertising from the media. Again, the economic impact of the ruling was significant. Cigarette advertising accounted for $216 million in television receipts in 1967—8 percent of all 1967 television revenues from advertising. Cigarette advertising on radio amounted to 5.9 percent of all radio advertising, and totaled $17 million.

Congress acted, after much delay and lobbying by various interest groups, with the Public Health Smoking Act of 1969, which stated that:

It shall be unlawful for any person to manufacture, import, or package for sale or distribution within the United States any cigarettes the package of which fails to bear the following statement: "Warning: The Surgeon General Has Determined That Cigarette Smoking Is Dangerous To Your Health."[19]

The preemption that no federal, state, or local authority could require any other health statement on packages except the one required by the Act continued in force. The Act also stated that:

After January 1, 1971, it shall be unlawful to advertise cigarettes on any medium of electronic communication subject to the jurisdiction of the Federal Communications Commission.

During the course of Congressional consideration of the Act, the cigarette manufacturers and networks reached voluntary agreement to phase out all cigarette advertising by January 1, 1971.

Programs in Other Countries. Several countries prohibit or restrict cigarette advertising on television and/or radio. Cigarette advertising is barred from television in England and France, and from television and radio in Czechoslovakia, Switzerland, Canada, and Italy. Ireland has banned completely all cigarette advertising in electronic and print media.

Increased attention also is being given in a number of countries to the production and promotion of cigarettes which are low in tar and nicotine. Cigarettes with tar and nicotine levels "well below average" now account for 75 percent of total cigarette production in Austria. Cigarettes are tested for tar and nicotine content in Sweden and Canada, and the results are made available to the public.

The English experience after cigarette advertising was banned on British television in 1965 is of particular interest. In a study on the effects of the television advertising ban the British social scientist John Wakefield made the following points:[20]

(a) the immediate loss of revenue by commercial television stations was made up in the first year by increased revenue from advertisers of other products;

(b) there was a switch of cigarette advertising from television to print and an increase in the amount spent on advertising of cigars and pipe tobacco in all media;

(c) promotional expenditures on gift coupon schemes doubled in the first year after the television ban;

(d) the decline in cigarette tobacco consumption by men since 1960 continued in 1965 and 1966, but consumption by women rose in 1966 after falling since 1963;

(e) the proportion of smokers who used only pipes or cigars rose in all age groups of the male population, including for the first time the 16-19 year olds.

413. Solutions to Major Internal Pollution

There are two distinct issues involved in a discussion of possible solutions for problems of internal pollution. For pollutants such as mercury and cadmium, the problems of cost internalization are analogous to those applicable to DDT, oil pollution, the SST, and nuclear wastes. Here, however, the potential scope of cost externalities is more limited than was the case for the SST, for example, and the kinds of economic and legal approaches to cost internalization discussed in the following chapters are applicable.

With substances such as cyclamates, drugs, and cigarettes which are potentially harmful, the problem is somewhat different. Here, it can be argued that the burden of potential health damage is voluntarily assumed by the consumer, and that he should have the freedom of choice to assume this risk in return for the satisfactions provided him by the product. However, in order for this voluntary risk-acceptance to work, the consumer must be informed of all the possible hazards to which he is exposing himself—there must be complete and full disclosure of all the detrimental side effects involved, including possible synergistic-carcinogenic effects. This happened to some extent with cyclamates, although the risk-acceptance became incidental when the product was removed from sale. The public has clearly not been informed of all the dangers inherent in the use of birth control pills or cigarettes. To the extent this has not been done there are some costs involved in the use of these products which have not been voluntarily accepted by consumers. For example, smokers in general have never been adequately informed of the relationship between pulmonary emphysema and cigarette smoking; it is probable that a number of sufferers from emphysema do not recognize their symptoms as calling for an immediate cessation of smoking.

One approach to full disclosure which may be of some use is that found in the fairness doctrine, which requires broadcast licensees to present an overall balanced presentation of controversial issues of public importance. For the fairness doctrine to be applicable, there must be a controversial issue involved, and there must not have been an overall balanced discussion of the problem on the particular station. The onesidedness of the message is determined not by studying each word in an advertisement or presentation, but by judging the ultimate impression upon the mind of the listener from the sum total of what is seen and heard, and from all that is reasonably implied. In the cigarette ruling the court recognized that although cigarette commercials never expressly stated that smoking was not harmful, the image created by the ads was that of persons enjoying normal healthy lives while smoking their cigarettes. Therefore the FCC and the court found an implicit assertion that smoking is a normal and nondangerous activity. If the general impression test were applied to a televised panel discussion which compared birth control pills to other available methods of contraception without any discussion of detrimental side effects, the same result might be reached.[21] If the fairness doctrine did apply, then broadcast licensees would have to donate time to respondents presenting the "other side" of the issue in controversy. To avoid this requirement of free time, and the resulting loss to the licensee both of saleable time and of advertiser goodwill, a licensee would certainly require considerably more full disclosure in the initial presentation of controversial ecological issues.

There is one other legal theory under which ecology groups may be able to buy advertising time to rebut one-sided presentations if their request for free time is denied. In the *Red Lion Broadcasting* case,[22] the Supreme Court recognized that the first amendment protects the collective rights of the viewing public to suitable access to social, political, aesthetic, and other ideas. This interpretation of the first amendment maintains that it is the right of viewers and listeners to hear, and not only the right of a particular speaker to speak, that is protected. Until the *Red Lion* decision, broadcasters had not been required to accept advertising dealing with controversial issues. Allowing ecology or similar groups a paid rebuttal to incomplete information about potentially dangerous products would be an inferior solution to allowing free rebuttal, but—given that such rebuttal were forthcoming even at a price—it would provide a first approximation to removing the external costs now implicit in the use of internal pollutants such as cigarettes and birth control pills.

REFERENCES

[1]Reported in E. L. Hofmann, *The Use of Benthic Communities in Pollution Studies,* unpublished paper, Wellesley College (June, 1971). The statistic has led to the novel speculation that a few hundred tons of oysters, planted alive in the right places, could be used to clean up a substantial portion of all the DDT in coastal waters of the United States.

[2]Samuel Fogel, Ilan Chet, and Ralph Mitchell, "The Ecological Significance of Bacterial Chemotaxis," a paper presented at the annual meeting of the American Society for Microbiology in Minneapolis, Minnesota (December, 1970).

[3]It is fair to say that some pests are a *creation* of the pesticide industry, which has released species from their natural restraints through killing off their predators. For example the emergence of the European red mite as a major pest in apple orchards followed the use of DDT to control the codling moth. In a similar way, pesticides may directly result in an increase in the population of their victims. Azodrin is a broad-spectrum insecticide which kills most of the insect populations in a field. Unlike the chlorinated hydrocarbons, it is not persistent; its effects are most devastating in populations of predatory insects. Therefore when a treated field is reinvaded by pests, their natural enemies are often absent. Experiments by University of California entomologists indicated that, rather than controlling bollworms, Azodrin applications through their effect on the bollworm's natural enemies, actually increased bollworm populations in treated fields.

[4]This is also the conclusion reached by the Mrak Commission, which considered the economic aspect of DDT at some length. See *Report of the Commission on Pesticides and Their Relationship to Environmental Health* (Washington, D.C.: Department of Health, Education, and Welfare, December, 1969).

[5]Typical of the undocumented estimates presented is one by H. L. Straube, general manager of the agricultural chemical division of the Stauffer Chemical Company, that if pesticides were withdrawn in the U.S., crop and livestock losses would approach $20 billion per year and the price of farm products would increase from 50 to 75 percent.

[6]Rachel Carson, *Silent Spring* (Boston: Houghton Mifflin Company, 1962).

[7]*Crowther* v. *Seaborg*, 312 F. Supp. 1205 (1970).

[8]There is no obvious logic in distinguishing a poison that causes cancer from one that produces mutations, liver disorders, or blood disease. Whether acceptable dosage levels can be established for some of these diseases and not others is an empirical issue, of course. The limitation of the Delaney amendment existed solely because Congress at the time of its passage seemed to feel that the unique characteristics of cancer justified its "separate consideration." In any event, consistency would seem to require that the Delaney Amendment be extended to all substances found to be mutagenic, teratonogenic, or to all potentially dysbiotic chemicals for which there is no known safe dosage level.

[9]In retrospective studies, data from the personal histories and medical and mortality records of individuals in groups are considered. In prospective studies, women are chosen randomly or from a special group such as a profession and are followed from the time of their entry into the study for an indefinite period, normally until they die or are lost to the study for other reasons.

[10]See Advisory Committee on Obstetrics and Gynecology, Food and Drug Administration, *FDA Report on the Oral Contraceptives* (Washington, D.C.: U.S. Government Printing Office, August 1, 1966).

[11]In the past, several manufacturers of birth control pills (notably Searle, the manufacturer of *Enovid*) have commented for publication that there is evidence of a lower incidence of cancer in women who take the pill than in non-users. This is probably true, but the statistic is misleading. Careful physicians examine women for evidence of tumors before prescribing the pill, and if they find cancer or certain other conditions, will not prescribe them.

[12]There is also an operative problem in finding appropriate samples of patients for study. In a study of the incidence of breast cancer with 4-year followup of women aged 20 to 39 years, a sample of 60,000 to 80,000 person years, or 15,000 to 20,000 women, would be required to have a 90 percent chance of detecting (at the 95 percent probability level) a simple doubling of risk. A control group of the same size would be required, and the number of dropouts from either group would have to either be minimal, or the actual dropouts carefully interviewed. Changes in the incidence of cervical cancer would require samples of about the same size, while changes in the incidence of endometrial cancer would require samples about seven times as large to obtain results as accurate as these for breast cancer—which means 130,000 women in the control group, and another 130,000 women in the experimental group. No studies even approaching this magnitude have ever been reported.

[13]*Report of the Advisory Committee to the Surgeon General Concerning Smoking and Health* (Washington, D.C.: U.S. Government Printing Office, January 11, 1964).

[14]The Council for Tobacco Research, *Health Effects According to the Tobacco Research Council, Report of the Scientific Director for 1966-1967* (New York, 1967), especially pages 4-14.

[15]The theory that predisposing factors in one's heredity make one more likely to contract a given disease has received fairly widespread support in the medical community. However, accepting the predisposition theory does not mean that cigarette smoking is not implicated in these diseases. Persons with predisposing factors in their heredity may have a high incidence of lung cancer; persons with predisposing factors who also smoke heavily may have a much higher incidence of lung cancer.

[16]See *Cigarette Labeling and Advertising,* Hearings before the Committee on Interstate and Foreign Commerce, House of Representatives, Part I, 91st Congress, 1st Session (1969), pp. 560-570.

[17]15 U.S.C. 1331-1338, 79 Stat. 282.

[18]The Commission's position was sustained by the U.S. Court of Appeals for the District of Columbia Circuit in *Banzhaf* v. *Federal Communications Commission*, 405 F. 2d 1082 (1968).

[19]Public Law 91-222, signed into law on April 1, 1970.

[20]Cited in *Cigarette Labeling and Advertising, op. cit.*, pp. 190-191.

[21]The whole fairness doctrine question is considerably more involved than a short discussion would indicate. An excellent discussion of the doctrine and its application to an ecology case, that of misleading advertising for Standard Oil of California's F-310 gasoline additive, is found in Alan F. Neckritz and Lawrence B. Ordower. "Ecological Pornography and the Mass Media," *Ecology Law Quarterly* 1 (1971), pp. 374-399.

[22]*Red Lion Broadcasting Co.* v. *FCC*, 395 U.S. 367 (1969).

5

Economic Approaches

To Cost

Internalization

As I looked into the future
Far as human eye could see,
Saw a vision of the world
And all the wonder that would be.

Saw the landscape filled with chimneys
Oceans brown with floating scars,
200 million belching Chevys
The urban planners drunk in bars.

500. Economic Externalities Revisited

We have continued to point out that the discharge of pollutants imposes on some members of society costs which are inadequately imputed to the sources of the pollution by free markets. The result is more pollution than is desirable from society's point of view. Some sectors of society benefit from the existence of externalities. Those who buy the goods produced, and those who own land, labor, or capital inputs that are used in producing these goods, all benefit by not having to include the external costs involved in their selling price. The presence of externalities also leads to a misallocation of productive resources. For example, a newsprint manufacturer may pollute a river and thereby increase the cost of water treatment for a glass producer further downstream. From society's point of view the newsprint firm has understated its costs and is producing and selling too much, while not enough glass is

158

being produced, because glass prices are overstated by the water-purification costs imposed by the newsprint effluent. Such a misallocation of productive resources—producing too much newsprint, not enough glass—constitutes economic inefficiency.

Economists have generally adopted the position that complete efficiency could be attained only if all external costs were somehow internalized to the firms that produced them. This approach was discussed and attacked in a 1960 article by R. H. Coase,[1] who argued that, given costless bargaining, efficient resource allocation could be achieved regardless of which party was required to bear the cost of the externality. According to the Coase argument, the absence of bargaining costs means that optimal resource allocation is independent of society's laws, or "starting rules"—whether incinerators have a right to pollute or nearby residents have the right to breathe clean air, or whether SSTs have the right to produce sonic booms or persons along the flight path the right to peace and quiet is irrelevant. Bargaining and side payments would produce the same end solution, independent of which position or assumption one started from, and there would be no increase in human satisfaction stemming from judicially-imposed liability.[2]

The important assumption here is that bargaining is costless, that there are no mediators, arbitrators, researchers, or others to reduce the amount of money transferred among the parties to the bargaining process.[3] Because the actual cost of bargaining is not zero and can be very high, actual bargaining to arrange pollution rights will be undertaken only when the increased value of production upon the rearrangement is greater than the costs of bringing it about.

501. Potential Solutions to Externality Problems

Again and again, the imperfect working of the free market is cited to justify almost any kind of governmental intervention in the free market. In the case of air and water pollution, the existence of external costs often leads people to think entirely in terms of direct regulation in the form of permits, registration, licenses, or administrative standards. These approaches have not proven very effective means of controlling environmental pollution. Control by means of regulation offers the polluter only the crudest form of economic incentive not to pollute—the only exception being where fines imposed for violations exceed the cost of compliance. Regulation may also turn out to be economically dysfunctional; for example, a strict application of a rule that "all wastes amenable to treatment must be treated" might result in large increases in production costs without producing any corresponding benefit to the environment. Also, any impetus for research and development of better effluent controls is missing so long as minimum legal pollution standards are being satisfied. Even where the marginal costs of additional control efforts are less than the marginal gains which would accrue to society, a polluter has no economic incentive to take these efforts under a regulatory scheme.

Given these defects in direct regulation schemes, it makes much more sense to investigate how externalities may be reduced or eliminated through restructuring of badly-functioning market mechanisms. Only when such restructuring appears totally impossible should the advantages of free decision-making be given up in favor of direct regulation.

The "Do Nothing" Alternative. A first approach to dealing with externalities is simply to ignore them. This is economically sound if the bargaining or transaction costs of any internalization scheme exceed the resultant gains in efficiency. A failure of private bargaining to arise may indicate that there are such high transaction costs as to preclude any increase in efficiency through cost internalization. It has been observed that: "The absence of a price [or a market] does not imply that either market transactions or substitute government services are desirable."[4] If private bargaining costs are too large to bring the externality into the price-bargaining system, then judicially or administratively imposed pricing or regulation might also be too costly to justify in terms of any possible efficiency gain. Thus, for example, emission control standards or control devices on outboard motors might be neither economic nor desirable no matter how imposed.

Zoning. The costs of administrative ordering may be so small relative to private bargaining costs that the private market should be displaced entirely, as is the case with emission-standard zoning. One problem is that while the expense of internalizing pollution costs can be estimated from existing agency expenditures, costs of taxation systems, or costs of litigation, the potential gains from controlling any one form of pollution are more difficult to measure. Thus, justification for either regulation or for administrative ordering must rest on either intuitive cost estimates, or on an *a priori* criterion of equity that does not require a cost-benefit justification.

Bargaining. If bargaining is economically feasible, it is the simplest way by which economic allocative efficiency can be achieved in a simple world of one polluter and one victim. Prior rights are assigned by a legislature or by the courts, to the interests which are considered most important—the creation of a market by rights-creation.[5] If the right to use scarce air and water resources belongs to the polluter, the victim can seek a contractual arrangement whereby he will pay the polluter to abate his pollution, assuming, of course, that the value of the reduction is worth more to him than the payment, and that it is cheaper to control pollution at its source. If the cost of pollution reduction becomes too high, the victim would probably choose to endure the pollution. From the standpoint of economic efficiency this would be acceptable as the value stemming from pollution control would then be less than the cost of resources required for control. If it were less expensive to correct the results of pollution where they are felt rather than at the source, as may be true with some forms of water pollution, the victim would prefer to clean up his own environment. This would also be economically efficient.

The same economic result is reached if assignment of prior rights dictates that the use of the environment belongs to the victim, and that the polluter must pay the costs of his pollution. Polluters would pay for the right to pollute up to the point where additional payments are greater than initial control costs. The victims would accept these payments up to such point as the payment failed to compensate for the additional costs incurred. Either way, cost internalization would result, with its burden being dependent on the assignment of prior rights.

However, this and most other bargaining models postulate a simple framework of one polluter, one victim, and are thus limited in value. Even where the one polluter-one victim conditions hold, it is unlikely that most single victims would have the resources to adequately bribe a large polluter to cease pollution. If several victims are involved, the freeloader effect will provide disincentives to a bargaining solution. Since air or water are common resources, one victim's payment will produce pollution reduction advantages which also accrue to his fellow victims. Each victim is motivated to do nothing and let others do the bribing. The resultant lack of incentive among all victims prevents any action at all. If it is the victim who has rights to pollution-free water or air, bargaining enforcement may be equally difficult because the victim rarely has the resources to bring a suit to enforce his rights—and knowing this, the polluter has little incentive to negotiate a settlement with any individual victim.

Extending the Firm. A perfectly valid approach to the internalization of social costs is to enlarge the size and scope of the firm so that all the costs and benefits of externalities accrue to the same entity. This might result in such a vast expansion that the costs of underspecialization might be greater than the gain from internalization. Probably the only meaningful way to internalize the costs of air pollution, which normally are spread over a wide area, would be to invoke public ownership. As both economic sophistication and the degree of pollution increase, it is probably inevitable that externalities will become the subject of increasing state-controlled allocation.

Consider the earlier case study of the Delaware River Estuary. Extending the firm in this case would imply putting the entire estuary under a single management so as to internalize all benefits and costs. But although extending the firm would eliminate some external costs, technical externalities would remain in the production of water recreational facilities and the proper allocation of resources into recreational facilities by a free market would fail.[6]

Private Actions. Another approach to internalization of costs is the private civil action to enjoin or to collect damages from offending polluters. This approach enables the victim to obtain fairly rapid satisfaction rather than waiting for the implementation of legislation, or for appeals to regulatory bodies to be resolved. It is also flexible since the victim can direct his action against new and complex pollutants, or base his claim on new forms of pollution damage. However, problems of proof, excessive costs, and lack of judicial precedent are barriers facing the private litigant. The topic of internalization of costs by private actions is an involved one, and is treated in more detail in Chapter Six.

Direct Taxation. It is sometimes argued that a direct

tax on pollutants can serve as a surrogate for costless bargaining. A direct tax does allow firms to maintain some degree of flexibility in finding the most efficient way to minimize their pollution, and hence minimize their tax load.

However, cost-incorporating taxes are subject to a number of criticisms. It is generally desirable to minimize the degree to which taxation is used to control behavior rather than raise revenue. It is also argued that a tax on one person, based on the cost or damage to another, will always produce an unstable equilibrium.[7] It is not clear how one would design a tax system which, if based on pollutants emitted, would not be subject to locational bias and arbitrariness. Finally, once an estimate of the total cost of pollution is made, the cost must be apportioned among polluters. However, the actual contribution of any one firm will depend on interactions among pollutants, on the direction of prevailing winds, on smokestack height, on timing of emissions during the day and by season, and so forth. Any equitable tax would thus have to encompass so many variables as to be unworkable, or would itself produce locational and other diseconomies through its application.

The desirability of a direct tax on pollutants is also relative to the extent to which the initial distribution of income in society is inequitable. If everyone had the same income, it would be equitable to tax persons who used goods that lead to pollution of the environment, either to pay the costs of restoring environmental quality or to compensate others for the damage caused them. But in a world where incomes are not equally distributed, such a tax leads to material goods (which comprise a relatively larger part of the budgets of the poor) becoming more expensive relative to services (which comprise a larger part in the budgets of the rich). This means the taxation of staple goods used by the poor to pay for protection of the recreational amenities of the rich.

Equitable or not, such taxes are being proposed. The Council on Environmental Quality has suggested a penalty tax on the sulphur content of coal, oil, and natural gas to go into effect in 1974.[8] Under the proposal, the tax would be imposed initially at a level of one cent per pound of sulphur content of coal burned in the first year, rising to 10 cents per pound by 1976. The objective would be to provide time and the incentive for conversion to the use of low-sulphur fuel and for adoption of methods for burning high-sulphur fuels without discharging sulphur dioxide into the atmosphere. The tax would be imposed on the producers of fossil fuels; a user of high-sulphur fuel who burned it without discharging sulphur dioxide into the air would receive a rebate of the tax that had been paid on the high-sulphur fuel.

One problem is that more than half of the sulphur-bearing fossil fuels used in the United States are consumed by public utilities, whose prices and profit levels are set by regulatory bodies that permit utilities to pass on to the customer all their costs of doing business. The utilities thus have no incentive to do research to find more economical ways of eliminating sulphur dioxide emissions. However, there is not enough low-sulphur fuel in the United States, regardless of costs, to meet the needs of utilities and other fossil-fuel users. Thus, the one real hope of eliminating sulphur dioxide emissions is improved technology in the use of these fuels.

E. S. Mills has recently proposed a scheme for direct taxation of pollutants that recognizes the interrelations of air pollution, water pollution, solid waste disposal, sewage, plastic containers, and the other paraphernalia of modern technological life.[9] His proposal is that the government collect a materials-use fee on specified materials at the time they are removed from the environment by the original producer, or are imported. The fee for each material would be set to equal the social cost to the environment if the material were eventually used in the most harmful possible way. The fee would be refunded to anyone who would certify that he had disposed of the material, with the size of the refund depending on the method of disposal. A full refund would be given for re-cycled materials; and ecologically harmless disposal would earn a large refund; disposal in the most harmful way would earn no refund at all.

The economic advantage of such a scheme is that whenever two or more materials could serve the same purpose, for example biodegradable and nonbiodegradable materials for containers, the fees would make their prices reflect social costs, including disposal, rather than merely private costs, and the original choices of materials would come nearer to being socially optimal. The combination of the schedule of refunds and direct costs would provide an accurate guide to individuals in choosing a method of disposal. Administratively, such a scheme avoids the problem of monitoring the disposal of goods by measuring the amount removed from the earth by the first producer, a much simpler problem. The burden of proof is placed on the individual and not on the pollution control agency. It is probable that specialized firms would arise to perform disposal services and provide certification of the method of disposal in order to earn the available refund for themselves and their clients.

There are some difficulties with Mills' scheme. It would have to apply over a wide geographic area, otherwise one location would be making refunds to those who disposed of materials that had paid the fee elsewhere. There would have to be some sort of refund for materials incorporated in very durable objects like buildings or dams (which might be considered as forms of harmless disposal rather than use).

Most importantly, there would be huge administrative costs involved in setting fees and certifying refunds. But Mills' scheme does provide a first approach to the correct problem, that of global materials usage, and it does so by utilizing the price system rather than administrative fiat to correct the divergence between private and social costs.

Subsidies, Tax Credits, and Effluent Fees. The need to internalize costs does not necessarily require that producers bear the total burden. Since the public benefits both from clean air and a productive industrial sector, it is sometimes argued that the public should pay some part of the cost of controlling pollution. A simple form of economic incentive to abate pollution would take the form of subsidy payments to stimulate reduction of emissions over the long run. Subsidies might be geared to a percentage reduction from total potential emissions, to an absolute reduction, or to the attainment of an emission standard set by a government regulation. A subsidy system can be thought of as equivalent to a tax, to be utilized with external benefits in the same way that a direct tax is utilized with external costs. A second economic alternative is one which provides tax credits for capital investment in pollution abatement facilities and accelerated depreciation for such equipment.

A third alternative often used in conjunction with subsidy or tax credit arrangements is the imposition of an effluent fee system, under which the polluter is made to bear the costs of his disposal directly. A schedule of emission fees predicated on the amount of damage done to the environment is applied to wastes discharged into the atmosphere or the water. Charges can also be imposed as a purely punitive measure without any relation to the damages actually done, or to the costs of treatment.

The three alternatives of subsidies, tax credits, and effluent fees are by far the most advocated and most widely practiced economic approaches to cost internalization. Since each is complex in design and impact, each is discussed separately below.

SUBSIDY PAYMENTS

502. Introduction

One argument frequently advanced in support of subsidies to industry for pollution control equipment is that the assets employed by industry in their abatement programs are economically unproductive; they neither add to revenues, nor decrease costs. A firm has little incentive to buy a device that does not help to produce salable products, or reduce production costs— even when the government offers to pay part of the cost. However, it sometimes happens that control devices enable the polluter to recover wastes with some economic value. To the extent that economic waste recovery is possible, firms may be induced to install pollution control devices by payment of only part of the cost by government.

It is also sometimes argued that cash payments for pollution abatement measures are to be preferred to tax credits because of ease of administration, because grants allow the government to avoid superimposing new loopholes in existing tax codes, and because grants enable preference to be given to smaller, financially weaker enterprises that are most in need of assistance in implementing pollution control.[10]

Subsidies, in the form of performance payments for pollution abatement rather than partial payments of capital expenditures, provide a continuing incentive to reduce emissions to the most economic level consistent with the level of subsidy payments. In practice, however, payments for reduction in discharge levels require an estimate of what the magnitude of pollutants might have been without any control devices, or any subsidy. The polluter has an obvious incentive to exaggerate the quantity of pollutants he would have discharged. As Edwin S. Mills notes:

> The trouble is precisely that which agricultural policy meets when it tries to pay farmers to reduce their crops. Jokes about farmers

deciding to double the amount of corn not produced this year capture the essence of the problem.[11]

Under a system of rewarding actual waste reduction, polluters would have an incentive to adopt, or at least to threaten to adopt, processes which produce a maximum amount of waste, in order to be able to collect a maximum of payments for restricting waste discharges. In equity, payments would probably have to be made on a continuing basis to firms which moved geographically as a means of reducing waste discharges in one local area. At one extreme, a case might be made for payments to potential dischargers who refrained from going into business—e.g., from locating anywhere—because of the payment for nonpollution. Moreover, subsidies could lead to higher net profits in pollution intensive industries, and perhaps produce a socially undesirable expansion of those industries.

It is questionable also whether accurate measurement of most emissions, and thus of emission reductions, is possible given our existing technology. It is certainly doubtful whether engineering estimates of emission volumes would satisfy the legal requirements of proof which an emission reduction subsidy system would impose.

An appropriate criticism of capital grant subsidies is that they are often incentives to undertake inefficient means of waste disposal. Changes in production techniques or in the fuel burned frequently are more efficient in reducing waste emissions than are end-of-the-line facilities. Most of the subsidy programs which have been enacted to date are limited to incentives for capital investment and have the result of making capital expenditures artificially cheap in relation to process or fuel changes, or to curtailment operations until an episodic danger had passed. Because of this, the most effective means of pollution abatement may in actuality be discouraged by subsidy payments.

Another frequent criticism of subsidy schemes is that it is inequitable to pay someone to refrain from an act which he has no right to commit in the first place, and that to pay a polluter to cease imposing a cost on the remainder of society is a form of blackmail. There are two possible replies to such a criticism: when the entire community is a beneficiary, it is not necessarily unsound economics (or unsound social policy) to have the community pay a portion of the cost of the improvement. More significantly, society is now asking the polluter to abate a nuisance which in all likelihood he had no way of knowing would ever be considered as a nuisance at the time his plant or facility was built. Most industrial polluters are simply doing on a larger scale what industrialists have always done; it is society that has changed the rules, not the polluter who has changed his behavior.

Whatever the pros and cons of subsidy payments, direct federal grants for capital equipment related to pollution abatement have been growing rapidly. Capital grants made through the Environmental Protection Agency and relating to water quality, air pollution, solid waste disposal, pesticide regulation, and radiation standards, almost doubled from $670 million in 1971 to $1.3 billion in 1972. By far the largest slice of the pie goes to local waste treatment plant grants. The federal government makes grants to cities for sewage treatment plants and to some industrial operations, and covers from 30 to 55 percent of the total cost depending on whether the state involved also bears part of the cost.

TAX CREDITS

503. Introduction

The term "tax credit," or "tax subsidy," or "tax expenditure" is used to describe special provisions of federal or state income tax systems which grant tax exemptions to achieve various social or economic objectives. Special provisions may take the form of deductions, credits, exclusions, preferred rates, or deferrals. In most cases the government has the option of using direct grants, direct loans, interest subsidies, federal insurance, or a guarantee of private loans to achieve the same ends as do the tax subsidies. The tax subsidy concept views a deduction as an imputed collection of the tax that would have been due had the deduction not been available, with a simultaneous grant of funds by the government to the taxpayer in the amount of the tax saving.

Although tax credits to aid in pollution control are a relatively new development, credits to induce similar activity or behavior in the national interest have a long history. The investment tax credit was introduced to encourage the purchase of equipment and machinery; preferential tax treatment of qualified pension plans was introduced to foster broader pension plan coverage; the corporate surtax exemption was aimed at encouraging small business; the deduction of home mortgage interest from taxable income was an inducement to home ownership; and so on.[12]

Federal tax credits aimed at cleaning up the environment had their beginning in 1966 with the suspension of the 7 percent investment tax credit. At that time an exception was made so that the credit could continue for pollution control equipment. In 1968, the tax-exempt status of industrial development bonds was revoked, but an exception was made for bonds the proceeds of which went into pollution control facilities. The federal law containing the major specific tax incentive for pollution control was passed in 1969, when the Tax Reform Act added Section 169 to the Internal Revenue Code. That section permits rapid, sixty-month depreciation allowances for newly-installed pollution control equipment.

The exemptions and tax incentive were motivated by industry arguments similar to those advanced in support of federal subsidy payments: that many industries would find it difficult to meet the cost of federal, state, and local pollution regulations without an exemption or incentive; that pollution control facilities do not add to earnings, cut costs, or improve competitive position; and that investment in pollution control produces a social benefit and the public should bear some of the cost of producing this benefit.[13] In each instance the Treasury, the Department of Health, Education and Welfare, and other government bodies opposed this special treatment for reasons ranging from erosion of the effectiveness of federal fiscal policy to the allegation that tax credits simply are not an effective stimulus to pollution abatement.

Secretary Finch of HEW pointed out that the cost to industry of effective pollution control under the then-existing regulations would average less than one-third of one percent of value added by all manufacturing and electric power industries, and that this relatively small cost did not appear to warrant federal cost-sharing.[14] The Secretary also argued that the proposed tax incentives were only available for investment in end-of-the-line hardware, producing an incentive for businesses to use hardware as a solution to every pollution problem to the exclusion of methods such as changes in fuel, in processing techniques, or in raw materials utilization.

Despite all opposition the investment tax credit was passed. Under the five-year rapid-amortization provision a taxpayer could deduct the total cost of pollution abatement equipment in five years, even though normal tax rules would establish a longer useful life for the property.

Viewed as a tax measure, the Treasury estimated that equipment with a 50 year useful life would have received a tax benefit from the new rapid write-off provision equal to a 20 percent investment credit. Viewed as an expenditure provision, the House, in effect, proposed to appropriate $400 million annually to share costs for an effort that, from the evidence available, needed no subsidy, and for an approach which, in the view of the experts, would in the long run be ineffective and inefficient.[15]

Not all pollution abatement equipment is covered by Section 169. The five-year tax write-off is limited to only a "certified pollution control facility," that is, a separate identifiable treatment facility used to abate air or water pollution. A building does not qualify as a pollution control facility under the provision unless it is exclusively for treatment; it must not include any equipment which serves a function other than pollution control. Facilities which only diffuse pollution, such as a smokestack on a plant, are not eligible for rapid write-off. Most importantly, Section 169 does not provide for writing off the cost of fuel desulphurization facilities, or of other facilities to remove pollutants from fuel, because such expenditures cannot be separated from other income-producing activities of the enterprise.

Individual states have been far ahead of the federal government in offering tax credits for antipollution investments. Starting in the early 1960's, state tax incentives took the form of property tax abatements (currently available in 21 states), sales and use tax exemptions (in 13 states), and accelerated write-offs or special tax credit (in 8 and 5 states, respectively).

Some of the state plans have been highly unsuccessful. New York State has since 1966 provided an investment tax credit of one percent of the cost of constructing or improving facilities to control air pollution or treat waste. However only 34 tax returns filed over the first 5 years of the plan claimed the special deduction. Apparently most companies found it more profitable to continue the state's already liberal rules governing the write-off rate for industrial assets, rather than switch to the new tax-credit scheme.

504. Virtues and Defects

A large number of claimed virtues and defects have been imputed to tax subsidies. In light of the extensive use of these tax devices, some consideration of the more common arguments is warranted.[16] Many of the virtues claimed for tax subsidies are subtle ones, which relate more to the common usage of the tool than to its economic impact.

For example, tax subsidies are often promoted with the observation that they involve less government bureaucracy and less red tape than do other alternatives. While it is possible (although doubtful) that tax incentives have in the past been less complex than, say, direct grants, this is not an advantage inherent to the technique. It is not the tax device that makes a program simple, but the substantive decision to produce a simple program. Much of the promotion of tax incentive programs may be a reaction to badly-designed direct expenditure programs; a more creative solution would be to design better direct expenditure programs.

A variant of the bureaucracy argument is that social problems are best solved by private rather than public decision-making, and that tax subsidies promote private decision-making and thus should be preferred. The rebuttal is parallel to the previous one; just as a direct grant program could be designed to involve a minimum of bureaucracy, so could a program be designed to provide a minimum of government control and a maximum of private decision-making. A tax credit allowed for pollution control expenditure is often cited as a method of government assistance that would promote private decision-making flexibility, as the taxpayer and not the government would select the control method and the amount of money to be invested in it. However, an expenditure program under which the government matched on a no-questions-asked grant basis some portion of all expenditures on pollution control would equally preserve private decision-making.

A much more substantial criticism of tax subsidies focuses on their lack of equity. They are worth more to high income taxpayers than to low income taxpayers, and they do not benefit those who are outside the tax system because of low incomes or tax-exempt status. This criticism is valid for virtually all pollution tax subsidies, which in general were never carefully structured to be equitable. By way of example, under the present tax structure every corporation pays a tax of 22 percent of its taxable income and a surtax of 26 percent of its income over $25,000. Thus, an expenditure for pollution abatement facilities for a corporation in the higher bracket, under Section 169, is subsidized by the government at 48 percent of the cost, while a corporation in the lower bracket is subsidized at 22 percent, and a firm with no profits at all in a tax year gets no assistance, although its obligations in installing facilities may be the same.

The real financial problem in private sector pollution is the inability of small businesses to pay for control devices. Not only do tax subsidies provide higher grants to larger firms, but they benefit only those with capital to invest and income to be sheltered—by definition almost excluding the smaller firms in the economy. For example, a $1 million investment in an electrostatic precipitator, when depreciated in sixty months under Section 169, benefits fully only businesses with $200,000 in yearly profits to shelter from taxation. Businesses with less than this in profit either get little or no benefit, or cannot use Section 169. Even larger firms in need of assistance might not get it under existing tax subsidy provisions. The unprofitable Penn Central Railroad could not benefit from Section 169, although it certainly has substantial need for assistance to meet its antipollution responsibilities.

Such inequitable and irrational side effects are not restricted to tax subsidies for pollution control, but are common to tax subsidies as a class of incentives.

> Many tax incentives look, and are, highly irrational when phrased as direct expenditure programs structured the same way. It is doubtful that most of our tax incentives would ever have been introduced, let alone accepted, if so structured, and many would be laughed out of Congress. What HEW Secretary would propose a medical assistance program for the aged that cost $200 million, and under which $90 million would go to persons with incomes over $50,000, and only $8 million to persons with incomes under $5,000? The tax proposal to remove the 3 percent floor under the medical expense deductions of persons over 65 would have had just that effect . . . What HUD Secretary would suggest a housing rehabilitation subsidized loan program under which a wealthy person could borrow the funds at 3 percent interest but a poor person would have to pay 7 percent or 8 percent? That is the effect of the five-year amortization of rehabilitation expenditures contained in the recent Tax Reform Act.[17]

Aside from equity considerations, it is unlikely that most large polluters need public financial assistance to meet pollution control requirements. It is even more debatable whether we want to promote the downgrading and elimination of small business which is inevitable as the cost of required pollution controls rise, and government gives disproportionate aid to the largest and most profitable companies.

Tax subsidies are also inequitable because some of the tax benefits go to taxpayers for activities which they would have performed without the benefits, thus the subsidy stimulates no additional activity. This is the case where tax subsidies are given to assist in meeting pollution standards required by law. However, the criticism applies equally to a similarly-structured direct grant program. For example, direct grants for industrial sewage treatment may be given to firms that for legal or other reasons would have treated their effluent anyway.

Another problem is the burden which tax subsidies place on the federal budget. While overall limits may be placed on federal spending by Congress, it is almost impossible to apply such limits to tax subsidies, which once passed become noncontrollable federal outlays. If tax subsidies were structured as direct expenditures there would be no logical basis for such immunity; the sheltering takes place solely because of the device through which they were granted. The existence of tax incentives greatly decreases the ability of the government to maintain control over its expenditures and priorities, both as to the programs to be funded and as to amounts to be spent on particular programs and areas.

The design of a subsidy or the regulations governing it may produce results which are counterproductive to the intended goal. It has been pointed out that a tax credit for pollution control equipment focuses on expenditures for machinery as a control method, perhaps to the exclusion of more efficient control techniques. If the tax credit applied only to equipment and not to its operation, preference would be given to facilities with low operating costs, even if they required very large capital outlays. For example, a firm seeking to remove cigarette butts from cooling water prior to pumping it back into a waterway might have them removed by an unskilled worker with a tea strainer at a cost of $5,000 per year. Or, the company might purchase an elaborate machine for $60,000 and with annual maintenance costs of $2,500 to do the same job. With a 100 percent tax credit, which is sometimes recommended in Congress, the machine would cost the firm less.

A more subtle problem arises in that state and federal agencies are burdened by tax subsidies with the need to process thousands of applications for exemptions to the variety of state and federal taxes; by law, they must provide businessmen with detailed certification to meet exemption requirements of Section 169. This burden (which is obviously not unique to tax subsidies, but could arise under direct grants) produces either a reduction in the agency's other enforcement and monitoring activities, or a very cursory examination of exemption applications which makes abuses of the tax subsidy quite predictable.

Finally, tax incentives are unlikely to lead to minimizing the total amount of pollution because they are generally applied uniformly within the jurisdiction of the government unit imposing the tax. However, to grant tax incentives on the basis of political boundaries is to ignore that a dollar spent to prevent the discharge of waste into a body of air or water already polluted to the limit of its ability to cleanse itself yields much greater social returns than the same dollar spent to prevent discharges into more assimilative bodies.

505. Comparison of Rapid-Amortization to Other Incentives

In analyzing the economic impact of a rapid-amortization provision such as that found in Section 169, it is useful to restructure the provision to conform to competing tax and non-tax types of pollution control assistance. Paul R. McDaniel and Alan S. Kaplinsky have worked out comparisons of rapid-amortization to direct grants, interest-free loans, government guarantee of conventional financing, and investment credits; the material presented here is based on their calculations.[18]

Direct Grants. If Section 169 were reformulated as a direct federal grant program, a summary of the grant provisions would read as follows:

> Every corporation that purchases a $150,000 certified pollution control facility shall be eligible to receive a direct grant from the federal government on the following basis:
>
> —a corporation with profits exceeding $25,000 for the year is eligible for a grant of $11,952;
>
> —a corporation with profits under $25,000 for the year is eligible for a grant of $5,479;

—a corporation with no profits, or with a loss for the year, or which is nonprofit by nature, will receive no federal assistance;

—a corporation which follows pollution control policies not involving the purchase of pollution abatement hardware need not even apply.[19]

No elected legislature would vote a grant with such inequitable and irrational side effects, yet the United States Congress passed exactly such a measure in Section 169 of the Tax Reform Act of 1969.

An Interest-Free Loan. Section 169 can be viewed as providing an interest-free federal government loan in the amount of the difference between the taxes which would have been paid using regular depreciation on the pollution control facility, and the taxes actually paid under accelerated depreciation during the fiveyear write-off period. The loan is repaid after five years, when the corporation must forego depreciation deductions to which it would have been entitled had regular depreciation been taken. On a $150,000 pollution control facility, a corporation in the 48 percent tax bracket receives a loan that saves a total of $24,038 in interest, while the 22 percent tax bracket corporation saves only $11,029 in interest. An unprofitable corporation faced with purchasing a $150,000 facility receives no interest saving at all because it has no earnings. Such a firm must go to the regular commercial money market for funds with which to purchase the equipment.

Government-Guaranteed Financing. Section 169 can be formulated as a federal program which guarantees repayment of conventional corporate borrowing for pollution control purposes, thus securing a preferred interest rate for the borrower. On a full mortgage for the $150,000 expenditure in a market which normally charges 10 percent per annum, a corporation with more than $25,000 in profits receives the equivalent of a guaranteed loan at an interest rate of 7.98 percent. A corporation showing less than $25,000 in annual profits must pay the equivalent of an interest rate of 9.38 percent, while one with no profits or a loss must pay the full 10 percent rate of interest.

An Investment Credit. Section 169 can be reconstructed as an investment credit by determining the present value of the net tax saving resulting from using rapid amortization rather than conventional straight line depreciation. The rapid amortization provision is the equivalent of a 7.97 percent tax credit for investment in pollution control equipment for a corporation in the 48 percent tax bracket, a 3.65 percent tax credit for investment in the same facilities by a firm in the 22 percent bracket, and no tax credit for a firm which shows no profits. In each example a 10 percent imputed rate of interest is used, although results are comparable irrespective of the interest rate chosen.

Given the problems and the inequitable effects discussed above, it seems as a general rule that the burden of proof should fall on anyone proposing the use of a tax subsidy system for pollution control in any given situation. Proof must include a detailing of what advantages, if any, might be obtained by using the tax subsidy as compared to spending an equivalent amount of money on direct pollution grants, or comparable alternatives. It appears that the advantages of a tax system would have to be overwhelming to overcome the problems and loss of control that accompany even a well-designed tax subsidy system.

EFFLUENT FEES

506. Introduction

As we have already noted, much of our environmental degradation arises because the price system is not applied to many of our natural resources. Fresh air and clean water are resources that are converted in the productive process in the same way that coal and steel are converted. But while a price related to the cost of production is charged for fuel and raw materials, our air and water resources can in most cases be used without payment for the privilege.

The problem exists because people use costly materials with a high degree of efficiency, but apply very little care or diligence to the use of resources which are free. In New York City, where water is supplied by the city at no charge, taps are left running, water lines develop huge leaks and are not repaired, and a periodic water crisis is the usual result. For the same reason, the cleansing power of air and water is overused when no charge is made for these resources. Consider, for example, how inefficiently electricity or long distance telephone service would be used if they were available at no charge.

The economist responds to this problem by claiming that there is no excuse for supplying scarce resources free; that these resources should be available only at an appropriate price. Specifically, the economist calls for an extension of a tax or fee system, one that does not necessarily increase the overall burden of taxes but rather gives industry the opportunity to minimize its tax load by behaving in a way consistent with social goals.

For example, the accumulation of litter could be reduced by imposing a significant tax or fee on no-deposit, no-return containers, perhaps matched by a reduction in the excise tax on items in returnable containers. Such an approach has the virtue of being self-enforcing, and therefore not very costly. Its instrument is the production line meter rather than the regulatory agency inspector. Calculation of the total tax payable on disposable containers requires no more than a record of how many cases of such containers have been manufactured or used. There are no crimes to be discovered, no courtroom battles, and no disputes over appropriate levels of fines to be imposed. Such a tax/fee approach has the additional advantage of longevity. Because it is self-enforcing it will be equally effective five or ten years from now when public interest in the subject has diminished. Unlike a program dependent on the enthusiasm of a regulatory agency, a tax or fee does not require continued enthusiasm for the original cause.

An effluent fee system is analogous to this tax on containers in that it attempts to minimize the costs of pollution damage and pollution abatement by requiring a polluter to pay a periodic fee, based on the amount of his effluent. One approach is to set the fee to produce that amount of effluent yielding the minimum total cost of pollution plus cost of pollution control. This point is illustrated in Figure 5-1, which is the total cost of pollution and pollution control curve that appeared in Chapter Two. Operation at point 'a' motivates polluters as a group to use an optimal mix of abatement and fee payments to maximize their profits.

Alternatively, fees could be based on the cost of treating the discharged waste and returning it to its natural state. This is practical with liquid wastes, but impractical with air pollutants because of the difficulty of treating gaseous wastes centrally. The approach has proven promising for water pollutants, and has been used in the Ruhr basin Genossenschaften, and by some municipalities in the United States for determining sewage charges for industrial waste. Effluent fees might also be imposed at a purely punitive level without relation to either the cost of pollution or the cost of

FIGURE 5-1

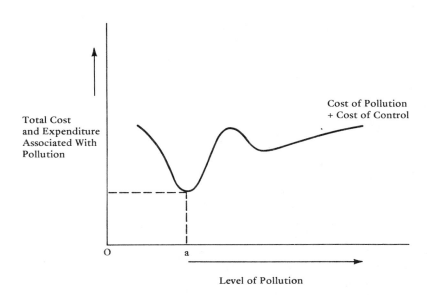

Level of Pollution

pollution control. However, this would induce individual polluters to undertake abatement to the left of the minimum total cost point on Figure 5-1.

Most commonly, however, effluent fees are based on some rough estimate of the average damage done to the environment and its members by a specific pollutant. For example, California state officials might conduct an investigation with respect to the Sacramento River basin, and conclude that 20,000 pounds of phosphates per month were being discharged into the river system by processing plants, and that the phosphates caused damage estimated at $50,000 per month to water supplies, navigation, individual firms, and public recreation along the 90 miles of waterway before the river flowed into the Pacific Ocean. The state government would then levy an excise tax of $2.50 per pound on the discharge of phosphates into the river. It would make similar calculations with respect to each of the other important pollutants in the waterway. Each polluter would be required to monitor his own discharges, and to calculate and pay a monthly fee reflecting its total polluting activity. Spot checks would be made on the operations of different polluters, much the same as government auditors make spot checks on the self-reporting activities of companies for income tax purposes.

Some polluting firms, aware of the technologies currently available, would purchase abatement equipment to reduce their tax burden to near zero. Other firms would find such equipment unavailable, and would respond by maintaining their existing production techniques but carrying a heavy emission fee burden. Some firms would choose to undertake some abatement and to pay some fees. In total, society would approach what was described in Chapter One as an optimal level of pollution; that level at which, to produce a dollar's worth of satisfactions in a less-polluted environment, it would be necessary to spend resources that were currently yielding more than a dollar's worth of social satisfactions in their current usage.

It should be noted that while the effluent fee system makes it unnecessary for an outside body to dictate to a polluter what abatement technology (if any) he should use, it does not remove the necessity of estimating the cost of the harm caused by specific effluents. The difficulties in such estimation have been discussed earlier. Theoretically, the effluent fee should be set equal to the cost of the marginal amount of harm done by the final unit of pollutant introduced into the environment. In practice, effluent fees would probably be set at the average level of harm done by a unit of effluent; marginal rates would be both difficult to calculate, and difficult to explain to the public. In most cases fees based on marginal harm would be higher than fees based on average harm, but the difference should not be so great as to invalidate the approach.[20] It should be recognized that the same average vs. marginal problem applies to the legislative approach to pollution control, although that approach is much more crude, and the differences involved much more significant than the average vs. marginal distinction associated with effluent fees.

It might also be possible to adapt the principles of two-level iterative planning as developed in Hungary and elsewhere to the specific problem of water pollution control. One such procedure requires the central authority to propose a scale of effluent fees to each polluter. Each polluter then makes his own cost calculations and reports back to the central authority the amount of effluent he will discharge and his total spending on abatement at each level of fees. Using this information, the central authority calculates a new schedule of fees, and the procedure continues until it converges to an optimal amount of pollution, and a least-cost combination of abatement to reach this optimal quantity.

507. Effluent Fees and Auto Emissions

A self-monitored effluent fee arrangement is best suited for stationary pollution sources producing significant amounts of effluent. Automobile emission fees would not be feasible if introduced at the owner level, because the monitoring of a large number of small-emission sources would introduce prohibitively high administration costs. A simple proposal is to sample the discharge of effluents from automobiles as they are manufactured, and to include in the sales tax on each car a fee based on the sampled effluent discharge. Sampling once at time of manufacture is inexpensive, but has the defect that it has no effect on cars once they leave the factory. Existing antipollution devices (blow-by devices, afterburners, and engine modifications) are not effective for more than one or two years without maintenance.

Thus, an effective application of effluent fees to automotive pollution would probably require both sampling of emissions at the time of annual inspection, and having the annual registration fee dependent on the result of the inspection. Ideally, one would charge a fee based on the average effluent per mile multiplied by the number of miles the car had been driven since its last inspection, and modified by whether the car was registered and driven within a metropolitan air shed.

A technological problem arises in that there is not currently available any inexpensive and reliable metering device for automobile effluents for use in an annual inspection. Federal automobile legislation has recognized these difficulties by avoiding any kind of fee system and opting for direct regulation at the manufacturer level.

508. Current Applications of Effluent Fees[21]

While effluent fee arrangements are only now coming into use in the United States, they have been in use in the Ruhr River basin in Germany for almost 60 years. The Ruhr and Emscher rivers flow approximately parallel to each other for about 75 miles in northwestern Germany until they both join the Rhine River near Duisburg. About 1904, users of the river basin area joined together and agreed to use the Emscher for the bulk of their effluent discharge, so that the Ruhr River would remain pure enough for recreational use. By 1930 seven Genossenschaften had been established in the Ruhr district, the Emscher River had been rerouted, diked, shortened, and about 90 percent of the liquid wastes from the area were being discharged into the Emscher rather than into the Ruhr. At four different points prior to where the Emscher empties into the Rhine, the Emschergenossenschaft operates treatment facilities to remove the wastes from its waters. The concentration of effluent into this single stream results in significant economies of scale in the eventual treatment of the waste.

Individual polluters discharging into the Emscher are charged effluent fees based on the quantity of their effluent, and on the average cost of treating each type of waste. The discharger thus is given an incentive to reduce his emissions to the extent that the cost of his own treatment is less than the charges levied by the Genossenschaft.[22] The possibility of saving on effluent fee costs has induced Ruhr industries to reduce the quantity of their emissions by instituting process changes and by recycling wastes.[23] Assessments on individual members of the Ruhr system are public obligations and can be enforced by law as taxes. Because of good relations between the associations and their members, there has never been a need for such enforcement. The Ruhr system is limited, as are the systems in use in the United States, to the abatement of water pollution. However, the method of determining discharge fees, if not the joint treatment concept, is applicable to the treatment of air pollutant discharges.

In the State of Vermont, the Water Resource Board has classified the waters of the state according to their intended uses and the minimum standards of effluents which may be discharged into each. A permit is required for each waste discharge. If the polluter is unable to meet normal standards required for issuance of a permit because his waste is not sufficiently treated, the Water Resources Board will issue a temporary permit but the discharger must pay an effluent charge based on the amount of damage to the waterway caused by his pollution. The charge per unit of pollutant varies to reflect the initial state of the waters involved, and the predicted economic damage done to other private and public users of the waterway. The Vermont charge system is not very close to the pure economic model of effluent fees discussed earlier, in that it merely superimposes an effluent charge on a more traditional permit system. However, the Assistant Attorney General of Vermont for Environmental Control has characterized the system as a necessary intermediate step in moving toward a system which relies exclusively on emission charges. An interesting interpretation of the Vermont statute is that the issuance of a temporary permit for discharge and the payment of effluent fees does not immunize the polluter from damages or injunctive relief from private suits, but the polluter is allowed to deduct any damages paid by him due to private suits from the effluent charge due the state.

The Regional Water Quality Act of 1970 introduced by Senator Proxmire comes closer to the pure economic model of emission charges. The proposed bill establishes a schedule of national effluent charges for all pollutants other than domestic sewage, with fees reflecting the quantity and quality of the waste discharged and the resulting damage to the quality of the waterway. The Water Quality Act would not preclude criminal or civil actions against the polluter, and would not allow him to deduct the cost of private damage actions from his effluent charges. An average effluent charge of 10 cents per pound of oxygen demanding material discharged (the figure recommended in the Delaware River Estuary study) would produce approximately $2 billion in revenue each year.[24] The revenue could be channelled to states or cities for the construction of waste treatment plants, or given to water management associations in the regions where the fees originated to further their abatement work.

509. Virtues and Defects of Effluent Fees

The biggest advantage of the effluent fee approach is that it is the least expensive method of producing a socially optimal level of pollution abatement.[25] The fee provides an inducement to the polluter to develop more efficient abatement techniques, or to install pollution control equipment. Pollution costs are completely internalized within the firm whether the polluter abates or pays the emissions fee if that is the least costly alternative. Pollution abatement equipment manufacturers are motivated to research new technology, knowing that markets for their devices are available if their costs can be kept below the known costs of emission fee payments. Further, determination of the most efficient method of abating a given firm's effluent is placed on corporate management, where it belongs, rather than in the hands of a regulatory agency. The corporation is not biased in choosing among abatement, process change, and fuel changes as is the case with tax incentive systems. The polluter is allowed to pay the effluent fee during a period of transition to new manufacturing processes, rather than being forced by sudden changes in the law to abate immediately.

A further advantage of cost internalization by effluent fees is the adjustment of consumer purchases in favor of lower cost (and lower fee-paying) producers. Depending on the ability of the polluter to pass on cost increases to his consumers, the demand for goods produced with minimal pollution should increase at the expense of high-pollution substitute goods. The reallocation of purchases results in a lower total pollution load for society as a whole.

There is great selectivity possible in the way in which effluent fees are applied. They are adjustable to the time of day, the season of the year, or the existence of special weather phenomena, such as atmospheric inversions. The entire fee schedule may be altered as our knowledge of the synergistic effects of certain combinations of pollutants grows. Polluters may be induced to locate in areas where their damage to the atmosphere or waterways is least, through altering the fees imposed for waste discharge in different geographic areas.

As compared with the least-cost program for the Delaware River Estuary, discussed in Chapter Three, the effluent fee program has the important efficiency advantage that it requires much less in the way of information and analytical refinement. The initial DECS report was sufficient to provide an estimate of the required fee, and changes could be made if responses to the charge indicated a need for adjustment. Since the DECS report did not consider the possibility of process change, the charge they recommended was probably too high. Under an effluent fee system the charge could be quickly adjusted downward when the problem was recognized.

As new technology developed, the effluent charge could be gradually reduced while the waterway standard was maintained, or the standard could be raised if this were considered desirable. The direct control measures implicit in the Delaware Estuary least-cost program (and of the effluent standard of the uniform treatment program) provide only a limited incentive to introduce new technology.

The argument most frequently cited by environmental enthusiasts opposed to effluent fees, and the one cited in the DECS report, is that the proposal creates a license to pollute. This suggests that the critic has in mind some alternative system of control that would both flatly prohibit pollution and would be economic in terms of alternative uses of available resources. Actually, the regulatory approach confers a much greater license to pollute. A regulatory order asks that pollution levels be reduced to some point which a government official finds practicable and feasible. When confronted with such an order, a firm has no incentive to improve its level of performance from that stated in the order. Until the order is issued, the firm has no incentive to do anything at all to reduce its effluent.

It is important to note that an insistence on the use of a regulatory approach instead of effluent fees does not imply the lack of a cost-benefit analysis (if only an implicit one). The assumption behind a regulation which bans all production or all pollution is that the costs involved are either infinite, or at least so much greater than potential benefits that the calculation is not worth making. This is certainly the case with the banning of dumping of mercury into lakes and streams, which was mentioned in the Preface.

There are two significant technical problems which must be overcome if effluent fee systems are to function efficiently. One is that current monitoring technology has not yet produced efficient, low-cost devices to measure all types of pollutants. Instruments capable of measuring several pollutants simultaneously are available, but their high cost makes it unlikely that they would be applied to the measurement of small-scale pollution sources. The problem is alleviated but not solved by the introduction of a self-monitoring scheme carried out by individual polluters, because this simply passes the high cost of available control devices on to industry, and says nothing about the simple unavailability of devices to measure a number of important pollutants.

The second technical problem is that of determining and quantifying the damage done by particular pollutants—a problem which has been discussed at length earlier. This problem is important because substantive due process requirements in law require that the cost-benefit relationship which forms the basis for the effluent fee schedule be reasonably accurate. There is some possibility that a polluter could challenge the validity of an effluent fee levy on the grounds that its fee calculations lacked scientific substantiation. The issue of legal acceptability of a cost-benefit analysis would not arise in an effluent fee scheme employing either the effluent standard-fee or the ambient air quality standard-fee approaches, which are discussed below. Under each of these two methods the fee schedule is prepared in relation to predetermined national standards rather than being based on estimates of pollution damages for a given area.

A further alleged weakness of effluent fee arrangements concerns the fact that some firms can pass the fee on to customers, thus eliminating their incentive to abate. It is not clear whether the entire burden of an effluent fee would or could be shifted to the consumer. The fact that effluent fees would be lower for rural polluters than for urban polluters, and lower for some competitors than for others, lessens the likelihood of prices simply being increased to cover the entire cost increase. The argument has greatest relevance when applied to public utilities, but even there the existence of a lag in obtaining regulatory rate-increases creates at least a short-term incentive for abatement.

The high degree of selectivity and flexibility inherent in waste discharge fees is sometimes viewed as a disadvantage as well as an advantage. Since the fee schedule is subject to continuous and rapid revision by pollution control agencies, industry may be frustrated in attempts to plan comprehensive abatement programs over the long term, and may be subjected to a rule of politically motivated men rather than one of equity. However, administrative controls could be introduced to guarantee against unjustified modifications, perhaps by guaranteeing existing fee rates which would not exceed stated limits within a given planning period.

In summary, the overriding advantages of emission charges are that when fees are imposed, all costs of pollution are internalized to the polluter, pollution abatement expenditures take on an economic rather than a social function, and flaws in the incentive-to-abate system which are common to other alternatives are corrected. Also, decision-making is decentralized, and individual polluters have very real incentives to adopt the most efficient means of abatement available. If fees are set at the proper level, a socially-optimal degree of abatement will take place. Six decades of German experience attest to the effectiveness of the effluent charge system; the several American experiments seem to point to a successful implementation of such systems, at least as applied to water pollution problems.

510. Administration of the Fee

The effective administration of an effluent fee plan requires an administrative agency to determine the cost-benefit relationship for all pollutants within its jurisdiction, to regulate and inspect self-monitoring reporting arrangements, and to calculate and enforce the resulting fee assessments. A question arises as to whether state, local, or airshed/watershed entities are best able to set up and administer such an effluent fee program.

The Air Quality Act of 1969 delegated all air pollution control responsibilities to state agencies. If an effluent fee plan could be implemented under this Act,[26] state legislatures would be responsible both for the enabling legislation and for supervision of control efforts.

Past experience with control efforts at various governmental levels indicates that local governments are probably better able than state governments to calculate the necessary cost-benefit relationships, to provide continuous monitoring and inspection, and to enforce fee payment. However, both local and state political entities are faced with a significant conflict of interest between their desire to protect the public welfare, and the economic reality that stringent pollution standards and high fee levels may induce new and existing industries to locate in more favorable jurisdictions.

A better approach might be to set up special pollution districts, conforming to airshed or watershed characteristics, to administer effluent fee control arrangements. These districts would almost certainly overlap existing city or county boundaries, and in many cases would encompass parts of several states. The pollution district would be identical with the area defined as appropriate for cost-benefit studies. The independent nature of the special district would make it less subject to pressure from large industrial polluters or local special interest groups. If such an arrangement produced overwhelming inter-jurisdictional disputes, it could be modified to one whereby the rule-making and fee-setting power of the pollution district authority was subject to challenge by city or state authorities in the courts on either procedural or substantive grounds. The authority would have primary responsibility for performing the initial cost-benefit studies and the setting of fees, for arranging for self-monitoring reporting systems, and for conducting economic studies of the business reaction to and economic impact of the various fees imposed.

The principal technical problem facing a control agency would be that of implementing a self-reporting monitoring program for the volume and nature of pollutants from individual sources. Ideally the agency would either provide each emitter with a standardized monitoring device to insure uniformity and accuracy, or it would certify available devices meeting agency standards. Until the necessary automatic monitoring devices were available at reasonable cost, the agency could require periodic samples to be taken and analyzed in lieu of continuous monitoring.

An administrative decision would have to be made as to whether the existence of an operating effluent fee plan should preclude private actions against polluters, or whether damages from such private actions should be deductable from effluent fee payments. Considerations of equity suggest that neither should be the case. A citizen subjected to pollution damage to his property or person would find little solace and no compensation in the increased fee that the polluter would pay to some autonomous agency. Further, if a quasi-monopolistic polluter such as an electric utility were actually able to pass much of its emission fee payments on to the consumer by way of higher prices, the individual citizen might find himself paying twice—for the emission fee, as well as suffering damage from the pollution. The retention of private as well as effluent-fee remedies would provide additional economic incentive for a quasi-monopolistic polluter to internalize his costs of pollution. For a polluter in a competitive industry (and if the cost-benefit calculations of the effluent-fee setter were correct), the existence of double liability would induce the polluter to abate his level of pollution below that level which was economically optimal from the standpoint of the whole society.

No such efficiency problem arises with alternate public remedies such as injunctive relief, which should not be displaced by the implementation of an effluent fee arrangement. For example, periodic pollution incidents arising from short-term fuel substitution may result in extraordinarily high pollution levels over a short period of time which would not significantly increase an effluent fee charge that is based on an average emission level over a period of time. Such a situation would warrant injunctive or similar relief.

511. A Pollution Standard—Effluent Fee Approach

An effluent fee administered by a special pollution control district may be combined with air or water quality standards for the district, providing an abatement standard upon which the effluent fee schedule is based. Such an approach has several names: here it will be called a pollution standard-effluent fee approach.

The air or water quality standard could be considered as a maximum standard above which the emitter could not pollute. Such a maximum standard could be set at the highest level of pollution acceptable without serious health effects. The effluent fee schedule below this maximum standard would provide a continuing incentive for the polluter to abate to a lower level of emission.

Alternately, the effluent fee schedule could be sharply increased at levels above the air or water quality standard applied to the district, and lowered below that standard. This would create an important incentive to abate to the standard level, and a less powerful incentive to abate below that level. Such a plan abandons the basic "cost-internalization to produce efficiency" approach, since the polluter pays a fee unrelated to the actual damage caused. However, this combined approach retains most of the other advantages of the effluent fee. The pollution standard effluent fee approach may also be more politically acceptable than an effluent fee approach which relies completely on a profit maximization incentive.

512. Conclusion

It is obvious that solutions must be found that are more likely to prevent the continuing deterioration of the environment than those which have been tried to date. Given the nature of our capitalistic society, economic incentives seem to have the greatest chance of successfully internalizing the social costs of pollution to the polluting firm.

Under existing regulatory plans, no incentive is offered for a polluter to abate beyond the minimum requirements of the law; in fact there is an economic incentive to continue to pollute at the same level as one's competitors. The various tax incentive and subsidy schemes which we have discussed recognize the nonproductivity of pollution abatement equipment, but it is doubtful whether such programs actually provide any incentive to either pollution control or to the internalization of pollution costs. A more serious problem is the unintended impact and side effects of incentives delivered through the tax system.

The most promising approach we have discussed is the effluent fee system, which overcomes the basic flaw in tax incentive and subsidy systems in that pollution abatement expenditures for end-of-line treatment or for production changes take on an economic function. An equally important advantage to an effluent fee system is the decentralization of the abatement decision-making process so that individual decision-makers have real economic incentives to adopt the most efficient available means of emissions abatement.

Another significant method of forcing industry to internalize the social costs of their production processes is private environmental litigation—an approach which is discussed in detail in the next chapter.

REFERENCES

[1]R. H. Coase, "The Problem of Social Cost," *The Journal of Law and Economics*, Vol. III (October, 1960), pp. 1-44.

[2]By way of illustration, assume that an optimal mix between soldiers and civilians is to be determined in a free market, with bargaining and side-payments permitted. There are two alternatives: taxpayers can bargain inductees into the military with higher payments, or everyone is drafted with the possibility of the highest bidders buying their way out. Coase argues that efficiency is served in either case because the tax cost of recruiting soldiers with dollars is equivalent to the cost of foregone national income of refusing to let soldiers buy their way out. Coase also argues that an identical soldier-civilian mix will result under either rule, with the last person to volunteer under a bid-in system also being the last person to buy his way out under a property-rights system.

[3]There are other assumptions inherent in a discussion of a bargaining solution; a small number of parties, equal in economic power, and in full possession of information concerning their own and their adversaries' positions, which confront an externality situation in an economy

in which resource allocation is in every other respect optimal. In most environmental pollution cases the parties involved are far from equal insofar as organization, power, and information are concerned. The typical situation is one in which one or more sources of pollution, associated with a large economic interest, affect a large and diffuse group of victims whose individual interests are harmed relatively little. Moreover, there is often no signal to the victims that important values are being destroyed. Consider a hundred fishermen who are faced not with dead, floating fish, but with a decline in their catch. No one fisherman finds it worth his while to bargain (or knows with whom to bargain), or even to generate information. The fisherman also finds that the costs of organizing a group action are prohibitive. In such a context, bargaining does not occur.

[4]A. Demsetz, "Toward A Theory of Property Rights," *American Economic Review Papers and Proceedings*, No. 57 (1967), p. 14.

[5]Lawrence Tribe, "Legal Frameworks for the Assessment and Control of Technology," *Minerva* (April, 1971), pp. 88-89.

[6]See the technical discussion of public goods externalities and water recreation in Paul Davidson, F. Gerard Adams and Joseph Seneca, "The Social Value of Water Recreational Facilities Resulting from an Improvement in Water Quality: The Delaware Estuary," in Kneese and Bower, *Water Research* (Baltimore: Johns Hopkins Press, 1966), pp. 182-187.

[7]The argument appears in "On Divergence Between Social Cost and Private Cost," *Economica*, Vol. 30 (1963), p. 309.

[8]A similar proposal made by the Treasury in 1970 to tax the lead content of gasoline was opposed by the oil industry, and didn't even clear the first congressional hurdle, the House Ways and Means Committee. In part because the oil industry is less likely to fight the sulphur tax, it is considered to have a much better outlook for success.

[9]E. S. Mills, *User Fees and the Quality of the Environment* (in preparation), cited in Robert M. Solow, "The Economist's Approach to Pollution and Its Control," *Science*, No. 173 (August 6, 1971), p. 502.

[10]Argued in Kenneth R. Reed, "Economic Incentives for Pollution Abatement: Applying Theory to Practice," *Arizona Law Review* No. 12 (1970), pp. 517-518.

[11]"Economic Incentives in Air-Pollution Control," in Harold Wolozin, ed., *The Economics of Air Pollution* (New York: W. W. Norton and Company, 1966), p. 77.

[12]These are discussed in Stanley S. Surrey, "Tax Incentives As A Device for Implementing Government Policy: A Comparison with Direct Government Expenditures," *Harvard Law Review* No. 83 (February, 1970), pp. 706-713.

[13]See *Hearings on the President's Proposal to Repeal Investment Tax Credit and to Extend Surcharge and Certain Excise Tax Rates* (Ways and Means Committee, House of Representatives, 91st Congress, 1st Session), pp. 145 ff.

[14]A February, 1970 report by the Conference Board in New York indicated that industry's 1969 capital appropriations for air and water pollution control dropped 56.9 percent below the 1968 appropriation, a drop in pollution control investments from less than four-tenths of one percent of 1968 gross revenues to less than two-tenths of one percent for 1969. Cited in Arnold W. Reitze and Glenn Reitze, "Tax Incentives Don't Stop Pollution," *American Bar Association Journal* No. 57 (February, 1971), p. 131. One estimate is that General Motor's budget for direct pollution control is about $40 million annually, about .17 of one percent of gross sales. This figure is about one-sixth of G.M.'s annual advertising budget. Cited in John C. Esposito. *The Vanishing Air* (New York: Grossman Publishers, 1970), p. 243.

[15]Paul R. McDaniel and Alan S. Kaplinsky, "The Use of the Federal Income Tax System to Combat Air and Water Pollution: A Case Study in Tax Expenditures," *Boston College Industrial and Commercial Law Review*, Vol. XII (February, 1971), p. 354.

[16]For much of the following material I am indebted to the discussion in Stanley S. Surrey, "Tax Incentives As A Device for Implementing Government Policy: A Comparison with Direct Government Expenditures," *op. cit.*, pp. 15-35.

[17]*Ibid.*, pp. 723-24.

[18]Paul R. McDaniel and Alan S. Kaplinsky, "The Use of the Federal Income Tax System to Combat Air and Water Pollution: A Case Study in Tax Expenditures," *op. cit.*, pp. 360-66.

[19]*Ibid.*, pp. 360-61.

[20]Marc J. Roberts has pointed out that marginal cost pricing presents a unique pricing problem. Since a river basin authority is operating a multi-part system and has the flexibility to treat first where the cost of treatment is least, the basin authority's marginal costs of removing additional waste will almost always be increasing and hence higher than its average cost. If all firms were charged the authority's marginal cost of abatement, a surplus would be generated. The river basin authority could avoid the surplus through a two-part tariff which incorporated a *negative* fee plus a variable fee based on the marginal cost of abatement. The negative fee would be a rebate given to each firm based on a non-pollution-related factor such as the number of employees or the dollar output of the firm.

A related problem is what happens when a new firm enters the system, which under marginal pricing will increase charges to each existing member since the expanded volume of effluent will have to be treated more intensively. One solution is to charge new firms a connection fee that reflects the costs of expanding the treatment system to accept their wastes. The fee could be fixed at some multiple of the annual negative basic charge and paid by making the new firm ineligible for the negative flat fee for an appropriate number of years. See Marc J. Roberts, "River Basin Authorities: A National Solution to Water Pollution," *Harvard Law Review* No. 83 (1970), pp. 1549-1553.

[21]The material in this section is summarized from the discussion in Kenneth R. Reed, "Economic Incentives for Pollution Abatement: Applying Theory to Practice," *op. cit.*, pp. 534-540.

[22]The net cost to a polluter who treats his own discharge is the difference between waste treatment expenditures and the value of recovered, salable by-products. The polluter will only recover these by-products if he treats his own waste; the existence of recoverable by-products will in part determine whether he will abate his own discharges, or will use the Emscher bulk treatment plants and pay the fee levied.

[23]See Allen Kneese, *Water Quality Management by Regional Authorities in the Ruhr Area*, in Hearings Before the Subcommittee on Air and Water Pollution, Senate Committee on Public Works, 89th Congress, 1st Session, part 3, pp. 942 ff. (1965), or the earlier article under the same title in *Papers and Proceedings of the Regional Science Association*, Vol. II (1963).

[24]Quoted in 115 *Congressional Record* S 14973 (November 25, 1969), cited in Kenneth R. Reed, "Economic Incentives for Pollution Abatement: Applying Theory to Practice," *op. cit.*, p. 539. To the extent that introduction of an effluent fee reduced the amount of liquid waste this figure would obviously be lower.

[25]The fact that an effluent fee approach will be the least expensive method of producing *any* desired level of pollution abatement has several interesting properties. Unlike many of the propositions about prices in welfare economics, this result does not require a world of perfect competition. It applies equally to monopolists or to oligopolists or to small firms producing heterogeneous products. The proposition holds even if the firms involved are not simple profit maximizers, but instead maximize their growth, their total revenues, their share of market, or any combination of these or a variety of other goals. Further, the proposition holds whatever the set of output levels or level of pollution abatement the society desires. It does not, for example, prejudge whether society should or should not reduce the number of private passenger cars in the process of decreasing air pollution. The only assumption inherent in this claim for an effluent fee approach is that a firm seeks to produce its chosen set of outputs at a minimum (private) cost. A formal

discussion of the effluent charge as a least cost solution is found in Allen V. Kneese, "Environmental Pollution: Economics and Policy," *American Economic Review* (May, 1971), p. 158.

[26]An effluent fee plan probably would not fall within the Air Quality Act's definition of an acceptable state control method, which is defined as one which directly limits the amount of pollution emitted. Since the effluent fee does not limit the amount of pollution but rather provides economic incentives to reduce emission levels, it apparently is outside the scope of the Act.

6

Legal Approaches

To Cost

Internalization

What do you do when a municipality decides that the highest and best use of a mighty river is an open sewer? What do you do when the Army Corps of Engineers or the Bureau of Reclamation decides to drown the Grand Canyon or most of Central Alaska, or insists upon destroying the delicate ecological balance of an entire state like Florida? Just what can you do?

SUE THE BASTARDS!

Industries and government can ignore your protests, ignore your picket signs . . . . But no one in industry or government ignores the scrap of legal cap that begins:

YOU ARE HEREBY SUMMONED TO ANSWER THE ALLEGA-TIONS OF THE COMPLAINT ANNEXED HERETO WITHIN TWENTY DAYS OR JUDGMENT WILL BE TAKEN AGAINST YOU FOR THE RELIEF DEMANDED.

Victor Yannacone[1]

600. Introduction[2]

In 1308, an unfortunate citizen of the city of London was executed by order of King Edward I for violation of a Royal Proclamation prohibiting the burning of high-sulphur coal instead of honest English oak in furnaces. Since 1308 there has been a decline in the severity of public and legal sanctions applied against those violating environmental laws, but public concern for the quality of the environment is being felt increasingly in the court-room. Private citizens have brought suits against the federal gov-ernment, the states, and private industry as well as against the state and federal administrative agencies which are supposed to be protecting the environment. There are at least a hundred suits pending in federal and state courts at any given time which involve environmental or pollution issues. Conservation groups such as the Sierra Club and the Environmental Defense Fund have resorted to litigation as the most feasible means of halting potential environmental damage.[3]

The threat of being sued imposes on a polluter an expected "cost" of potential private damages, plus an expected "cost" of defending suits that will be induced by the fact of pollution. The expectation that such costs will be incurred acts as a spur to industry to apply its technology and management resources to reduce the social costs of pollution. Litigation offers another way in which the external costs of pollution can be internalized as a cost of production, either through damages paid—by the polluter buying off the plaintiff, or by the polluter ceasing his polluting operations. The polluter who anticipates litigation has a strong incentive to spend up to, but not more than, a dollar on pollution reduction for every dollar of expected claims plus expected legal costs likely to be levied against him. This is exactly the way we wish the polluter to behave, for it accomplishes the objective of internalizing pollution costs without forcing an administrative body to go through the difficult process of extracting from the polluter all he knows about alternative technologies by which his production process can be carried on.

If private suits proliferated, and if a large enough number of polluters were assessed damages, the pollution control industry would expand both its research and its capacity in response to demand for more efficient control devices. Eventually, the resulting internalization of pollution costs would tend to increase sales of nonpolluting industries as consumer demand, if at all elastic, shifted towards goods whose price did not include the surcharge of court-imposed damages.

In private suits against industrial polluters, plaintiffs usually seek injunctive relief—a court order prohibiting the activity which caused the pollution. Monetary damages may be requested along with the injunction. However, damages are not often requested alone in environmental cases because a damage award does not prevent continuation of the polluting practice, which in most cases is the prime target of the plaintiff.

Also, a legal action for damages alone may be difficult to sustain because of the problems of measuring actual damages, and of allocating the award fairly among the victims of the pollution. Except in rare cases such as poisoning by a single pollutant from a single source, it is often impossible to isolate the effects of one contaminating discharge from those of another. This is especially true for health claims because humans are subject to many kinds of contaminants for long periods of time, and as indicated earlier, contaminants in combination may have detrimental synergistic health effects not related to the individual pollutants involved.

Dollar damages may be hard to establish because much pollution injury is irreparable. Health damage, wildlife destruction, and generalized disruption of the ecosystem cannot be repaired by monetary damages. Further, even when a particular plaintiff is well-compensated for harm to himself and his property, damage awards ignore other individuals who are potential plaintiffs and who for various reasons have not been able to bring legal action.

Thus, when a damage action is brought, the court is faced with the problem of awarding damages in an amount that is subject to great dispute, and against a particular defendant who cannot equitably be singled out as the sole guilty party in causing the plaintiff's damages. The court is likely to be reluctant to make an award of damages under such circumstances. It will be less reluctant when the plaintiff requests injunctive relief, since it is easier to demonstrate that the polluter is causing some harm, and that the plaintiff has suffered some injury. Injunctive relief puts direct pressure on the polluter to change his industrial processes or to cease production. Injunctions oblige the polluter to find a suitable alternative technology, or to negotiate with the plaintiff for an agreement dissolving the injunction in exchange for a satisfactory payment, or to terminate the polluting part of the operation.

In practice, the extreme burden of proof that has been placed on the plaintiffs in environmental cases has prevented cases— through damages or injunctive relief—against all but the most blatant polluters. For example in the case of *Gerring* v. *Gerber* (1961), an injunction was denied because the odor from the defendant's cleaning establishment, while admittedly overpowering, was no more so than that which is expected from that type of business.[4] However, the growing amount of information about the harmful effects of various pollutants, and the growing public concern for a cleaner environment, suggest that a trend toward greater success in actions for damages or injunctions against polluters will emerge.[5]

The cases filed to date present a great diversity of legal theories ranging from constitutional claims to a pollution-free environment, to more conventional legal theories of nuisance, trespass, and negligence. This chapter will briefly discuss the diverse theories for environmental redress which have been put forward in various cases, the limitations to effective legal cost-internalization, and the real function of the legal approaches now in use. What follows is not an exhaustive examination of all the judicial approaches potentially available to the victims of pollution, but it does indicate both established and novel ways of looking at existing legal doctrines, and hopefully suggests some wholly new approaches to pollution abatement through the courts. The reader must appreciate, however, that many of the doctrines examined below are burdened by either the weight of precedent, or by traditional judicial reluctance to pioneer new and uncharted frontiers of the law.

CAUSES OF ACTION

601. Underlying Theory

Before a court can award damages or injunctive relief, it must first have applied an appropriate legal theory. Historically, an individual whose person or property has been adversely affected by the use to which his neighbor devoted his property has been able to sue for damages or injunctive relief on the theories of nuisance, negligence, or trespass. These traditional remedies, considered below, were originally tightly structured legal concepts. In this century they have merged to some extent, with the boundary areas between them becoming increasingly fuzzy.

The traditional remedies all require a balancing of conflicting interests, and the value framework against which judges have performed this balancing historically has been weighted against environmental protection.[6] The courts have viewed both utility and harm in economic terms, while economic externalities which extended over more than a small geographic area were generally ignored. The objective of court decisions was always to encourage industrial expansion and economic growth, even at the cost of environmental damage; the common law doctrines now have the encrustations of a century of such attitudes.[7] This judicial attitude is well expressed in a 1954 Pennsylvania case in which the judge stated: "one's bread is more important than the landscape or clear skies. Without smoke Pittsburgh would have remained a very pretty village."[8]

Substantive law as it now exists is geared to the proprietary lawsuit and not to the suit to protect geographically diffused environmental values. The prospects for rapid change in judicial attitudes are promising, but uncertain. An examination of the available causes of action illustrates the variety of difficulties encountered in environmental litigation under present circumstances.

602. Nuisance

Nuisance law has traditionally been divided into areas known as "public nuisance" and "private nuisance." Public nuisance is the doing of, or failure to do something which injures the health, safety, or morals of the public, or creates a substantial annoyance or injury to the public. Private nuisance is a civil wrong for disturbance of rights in land, specifically for the unreasonable use of property so as to substantially interfere with the use and enjoyment by another of his property. While public nuisance has historically been associated with the removal of brothels, gambling dens, and similar institutions, its definition would seem to cover a situation where the air or water is being debased. However, the cases that have concerned smoke, dust, and water pollution have produced the finding that a private individual cannot sue to enjoin a public nuisance. The suit must be brought by the state or a federal attorney general in the name of the people of the state.[9] In 1972, there were only six states with statutes permitting individuals to sue to enjoin particular kinds of public nuisances.

The same act of pollution can create both a public and a private nuisance. Pollution of a waterway which destroyed its fishing would constitute a private nuisance where the river crossed private property and a public nuisance where public property (and public fishing) was involved.

Air pollution was recognized as a private nuisance as early as 1611, when an English court granted an injunction and damages to a plaintiff whose air had been corrupted by the defendant's hog sty. The defendant was found to be committing a nuisance even though he argued that a hog house was necessary to his sustenance, and that one ought not to have so delicate a nose as to be offended by the smell of hogs.[10] The following selection of cases suggests some more current examples of private nuisance applications, and some typical (and conflicting) judicial evaluations of the opposing equities involved.

In the frequently cited *Ducktown Sulphur* case,[11] the plaintiffs requested damages because pollutant discharges from the smokestacks of Ducktown's sulphur mills made it impossible for them to harvest their crops, largely destroyed the timber on their properties, and prevented them "from using and enjoying their farms and homes as they did prior to the inauguration of these enterprises." The court recognized the serious consequences of the pollution, but refused to grant injunctive relief. The court argued that since it was not possible for Ducktown to operate with less-polluting effect or to move to another more remote location, an injunction would compel it to stop operating its plants, making the property practically worthless and causing 10,000 people to lose their jobs. The court stated that:

> In order to protect by injunction several small tracts of land aggregating in value less than $1,000, we are asked to destroy other property worth nearly $2,000,000, and wreck two great mining and manufacturing enterprises. . . . The result would be practically a confiscation of the property of the defendants for the benefit of the complainants—an appropriation without compensation. . . . In a case of conflicting rights, where neither party can enjoy his own without in some measure restricting the liberty of the other in the use of property, the law must make the best arrangement it can between the contending parties, with a view to preserving to each one the largest measure of liberty possible under the circumstances.

The court did allow the plaintiffs to re-sue for monetary damages which were ultimately awarded.

In the 1911 *Hulbert* case,[12] the court took quite a different approach to an air pollution case which required a balancing of opposing equities. In *Hulbert*, the plaintiffs sought an injunction to require the defendant to stop discharging cement dust. The court found that the dust pollution was severe and not capable of being "dissipated by the strongest winds, nor washed off through the action of the most protracted rains"; that the value of the plaintiff's citrus fruit was decreased, that the presence of the dust on the leaves of the trees made harvesting of the crop very difficult and expensive, and that the presence of the dust in plaintiff's homes made life less pleasant.

The defendant argued that he was doing all he could to keep dust from escaping from his cement plant, that damages were sufficient to compensate plaintiffs for their injury, and that the court must consider the size of the cement plant payroll and its economic importance to the community in reaching a decision. The Supreme Court of California ruled that an injunction should be granted. The opinion admitted that the hardship inflicted upon the company by an injunction would be much greater than that suffered by the plaintiffs if the nuisance were permitted to continue, but argued that:

> It is by protecting the most humble in his small estate . . . that the poor man is ultimately enabled to become a capitalist himself. If the smaller interest must yield to the larger, all small property rights, and all small and less important enterprises, industries, and pursuits would sooner or later be absorbed by the large, more powerful few.

Both the doctrinal evolution in the law of nuisance and the economic analysis of one judge are illustrated in the 1963 *Renken* case.[13] Renken was a fruit grower in Wasco, Oregon who claimed that emissions from the Harvey Aluminum plant, consisting of particulates and gases, including fluorides, were harming his fruit trees. In examining Renken's request for an injunction, the court took careful note of the physical structure and chemical operation of the Harvey plant, with emphasis on the arrangements for exhaust and fume control. The court concluded that it was quite feasible to install cell hoods and electrostatic precipitators which would remove the particulates which were not removed by existing controls, and that such controls would reduce or eliminate the damage to the plaintiff's orchard. The court ruled that:

> While the cost of the installations of these additional controls will be a substantial sum . . . such expenditures would not be so great as to substantially deprive defendant of the use of its property . . . . If necessary, the cost of installing adequate controls must be passed on to the ultimate consumer. The heavy cost of corrective devices is no reason why plaintiffs should stand by and suffer substantial damage. . . . The defendant will be required to install proper hoods around the cells and electrostatic precipitators within

one year of the date of the decree. Otherwise, an injunction will
[be issued as requested] by the plaintiffs.

In scrutinizing the evidence relating to the plant's equipment
and processes, the court applied standards derived "from the best
contemporary practice among qualified manufacturers." One com-
mentator has suggested that the next appropriate step in a nuisance
case might be to test the performance of the manufacturer not only
by the criteria of technology actually available and in use, but also
by the efforts he may or may not have made to develop new
alternate technologies or engineering designs.[14]

The court's comment that "if necessary, the cost of installing
adequate controls must be passed on to the ultimate consumer" is
interesting both in that it ignores economic considerations of competitive market conditions and elasticity of demand in the
aluminum industry, and in that it raises explicitly the evolving
social view that the burden of compensating victims of pollution
should be carried by the users of the product whose production
caused the pollution. In requiring that the cost of corrective devices,
however large, be borne by the manufacturer so long as it is not
so great "as to substantially deprive defendant of the use of its
property," the court indicates that its decision was limited only by
the point at which confiscation might arise—which in most pollution cases would give the court a huge latitude to require corrective
action by the polluter.

Two recent cases based upon a nuisance theory have shown that
there is still strong judicial unwillingness to do more than award
damages in the hope that this might indirectly induce abatement.
In the 1970 *Jost* case,[15] plaintiff farmers sued a power cooperative
for damages for injury to crops and diminution of the value of
their farmlands. The court rejected defendant's arguments that the
company's exercise of due care should defeat a nuisance charge,
and that the social utility of the offending industry must be
balanced against the harm done. However, the court refused a request for an injunction, and awarded only compensation through
damages to the plaintiffs. If Dairyland Power increased its pollution in the future, provision was made for an increase in damages
to the plaintiffs, but there was no requirement either for future
injunctive relief, or for any provision to abate the pollution.

A New York court in the much-cited *Boomer* case was even more
explicit in rejecting injunctive relief as too severe a remedy to

impose on industry.[16] Boomer, the plaintiff and a landowner, brought suit against the Atlantic Cement Company, seeking an injunction to restrain Atlantic from emitting dust and raw materials in the operation of its plant. The court acknowledged that the cement plant was a source of air pollution and vibration nuisance, but denied an injunction stating that more research was needed to ameliorate pollution from cement plants. The court claimed that it was not the judiciary's place to spur such research. The court in its decision argued that:

> [Atlantic] expended more than $40,000,000 in the erection of one of the largest and most modern cement plants in the world. The company installed at great expense the most efficient devices available to prevent the discharge of dust and polluted air into the atmosphere.

The court chose to award permanent damages to Boomer; one effect of this remedy is that it terminated the lawsuit, for plaintiffs are precluded from future recovery because the defendant, by the payment of permanent damages, obtains what is known as a "servitude on the land." Several commentators have pointed out that, by paying the property owners permanent compensation for their land, a private company was able to seize private property and lay waste to the neighborhood.[17]

It is notable in *Boomer* that the court specifically chose not to use the long-established equitable remedy of an injunction that takes effect at a future date, thus allowing the defendant the opportunity to remedy existing nuisances. The remedy was rejected by concluding that technological breakthroughs in pollution control equipment were unlikely to take place in the near future, and that if at the end of a short period the entire industry had not found a technical solution to air pollution, the court would be hard put to close down one cement plant while leaving others in operation. The court did not consider that given the existing state of cement production technology, it was far less expensive for Atlantic to pay damage claims than to innovate research for pollution control devices, and the existence of continued damage claims would remove the incentive for either cement manufacturers or firms outside the industry to innovate new techniques or equipment. The effect of the *Boomer* decision on earlier case precedent is unclear; the precedent may be limited to those cases where the pollution cannot be abated by the most advanced pollution control devices, as opposed to situations where remedies are available although only at substantial cost to the polluter.

To the vagaries of the law must be added the fact that a private party seeking effective relief on a nuisance theory is also faced with a series of technical obstacles. The fact that a nuisance may be termed "public" has already been mentioned. In such a case the plaintiff must show special injury— which generally means that the plaintiff's damage must be different in kind rather than simply in degree from the harm suffered by the general public.

If a polluter has been active for some time, the legal doctrine of prescriptive rights may come into effect. Under a statute of limitations, his right to pollute would become absolute in regard to the particular plaintiff. However, many courts are refusing to recognize such a right because the defendant cannot meet various burdens of proof.

Another rule which may act against a potential plaintiff is that of "coming to the nuisance"—one example of which is the case where an individual may be found to have assumed an annoyance by moving nearer to a polluter. Coming to the nuisance would certainly weigh heavily in the balancing process of a court intent on comparing the relative equities of arguments.

In summary, a number of obstacles to an effective nuisance suit are frequently added to the currently ambiguous direction of current legal decisions. While it is frequently argued that the doctrine has potential for growth, nuisance at present must be said to be of only marginal effectiveness as a device for internalizing the costs of pollution.

603. Trespass

A number of pollution suits have attempted to use the theory of trespass, which is an unprivileged entry of a person or object on land occupied by another. The plaintiff's problems of proof are less under trespass than under nuisance. To establish trespass one need only show an intentional and unauthorized entry onto the land, while to show nuisance one must prove a substantial and unreasonable interference with the enjoyment of the land.

However, the theory of trespass requires a "direct" physical entry by a person or object, and courts have had conflicting opinions as to whether entry of smoke, fumes, or particulates onto a plaintiff's land qualifies as an "object."

For example, in the 1959 *Martin* case,[18] a group of Oregon cattle ranchers near a Reynolds Aluminum plant claimed that their cattle were poisoned by ingesting fluorides which escaped from the plant, and that forage and water on their land had been contaminated. Reynolds did not contest the facts, but argued that the mere settling of fluoride deposits upon land was not sufficient to constitute trespass. The court ruled:

> If we look to the character of [what] is used in making an intrusion upon another's land we prefer to emphasize the object's energy or force rather than its size. Viewed in this way we define trespass as any intrusion which invades the possessor's protected interest, whether that intrusion is by visible or invisible pieces of matter or by energy which can be measured only by the mathematical language of the physicist.

> We are of the opinion, therefore, that the intrusion of the fluoride particulates in the present case constituted a trespass.

There is still much conflict of opinion however, about whether air pollution (or water pollution) constitutes an "object," and cases since *Martin* have produced mixed results.[19] Also, some courts have held that if an intervening force such as wind or water carried the contaminants onto the plaintiff's land, that the entry is not "direct."

In summary, the fact that only plaintiffs in close proximity to the polluter have a cause of action, the difficulty of pinpointing which among many sources of pollution did the damage, the court's tendency to balance equities in trespass cases as in nuisance cases, and the cost of litigation against huge corporations, each discourages the filing of environmental trespass suits, and makes the trespass doctrine of only marginal value for effective pollution control on any large scale.

604. Negligence

The third conventional legal theory which is applicable to pollution cases is that of negligence. A direct causal relationship must be shown between the plaintiff's injury and the defendant's negligence, for negligence to be accepted by a court. In the 1958 *Blakely* case,[20] the plaintiff was allowed damages for negligence against Greyhound bus lines, for brain and nerve damage she allegedly suffered from inhaling carbon monoxide while a passenger on a bus with a defective exhaust system.

In the *Martin* case,[21] mentioned earlier, members of the Martin family brought a negligence suit against Reynolds Metals Company in which they claimed personal injuries arising from the fluoride compounds escaping from the Reynolds plant. The court ruled that:

> When [Martin] proved the emanation of fluoride compounds from the [Reynolds] plant, and the injury suffered by him as a result thereof, he made out a primafacie case of negligence on the part of the defendant.

In spite of the problems involved, a plaintiff may wish to pursue a negligence theory because of the greater likelihood of obtaining punitive damages than if the case were argued on nuisance or trespass theories. All but four states allow punitive damages, and about fifteen states award them with considerable frequency. Traditionally, punitive damages are awarded where there is malice, fraudulent or evil motive, or willful or wanton disregard of the interests of others. There is also justification for punitive damages when compensatory damages alone, although compensating the injured party, do not deter the polluter from committing similar acts in the future.

To date, the general "standard of care" problem in negligence cases allows the courts to balance the utility of allowing continued pollution in light of the general economic health of an area. What may prove to be the most productive approach in using negligence in pollution cases has yet to be tried—an application of the nuisance doctrine from the *Renken* case in a negligence case. If accepted by the court, a polluter would be negligent under the *Renken* doctrine unless he used the best available pollution control devices irrespective of their cost, so long as the expense does not bankrupt him. As in *Renken*, an unanswered question is what approach a court should take when a polluter uses available pollution control devices which are inadequate, but declines to conduct or share the cost of research to develop more efficient methods of pollution control for his industry.

605. Products Liability

The doctrine of products liability maintains that there is an implied warranty running from a manufacturer to the ultimate purchaser and to others who might be expected to use or utilize the product or service, that the goods are (among other things) not unreasonably dangerous. The logic of products liability is to insure that the cost of injuries resulting from defective products is borne by the manufacturer of the products, rather than by the injured persons who are powerless to protect themselves. Thus products liability shifts the costs of injuries from users of a defective product back to the manufacturer. To establish a manufacturer's liability it is sufficient to prove that the plaintiff was injured while using the product in the way it was intended to be used, and that the injury was a result of a defect in design or in manufacture of which the user was not aware.

An illustration of the possible use of products liability in environmental protection cases is given in the "Los Angeles Smog Case," filed in 1969.[22] Two citizens of Los Angeles, described as "C. Jon Handy, a land investment banker, and William R. Bernstein, a law student," suing for themselves, the People of The County of Los Angeles, the four minor children of C. Jon Handy, and all minor children similarly situated, sued the four major automobile makers and a number of oil companies for $15 billion because of their alleged critical role in the creation of Los Angeles smog. Named as defendants were General Motors, Ford Motor Company, Chrysler Motors, American Motors, the Automobile Manufacturers Association, Standard Oil Company of New Jersey, Gulf Oil, Mobil Oil, Texaco, Shell Oil, the American Petroleum Institute, Inc., the Secretary of Health, Education and Welfare, and the Attorney General of the United States. Three counts were listed in the suit, the first of which asked for an injunction that would require the automobile companies to alter, modify, or change their conventional internal combustion engines so that they did not cause smog, and to recall all existing automobiles in order to alter their engines so that they did not cause smog. The oil companies would be ordered to refine a "clean" motor fuel, and to refrain from adding tetraethyl lead or similar damaging substances to their fuels. The suit suggests that the doctrine of strict products liability can be used to internalize what are now costs external to the defendant companies, and in a way not possible through an action for nuisance.

Since the case has not yet been decided and there are no comparable cases known to the author, its implications can only be a subject of speculation. To continue their case, the plaintiffs will first have to satisfy some standard conditions appropriate to liability cases. They will have to demonstrate that automobile exhaust is the major factor in the causation of smog in Los Angeles, and that they have suffered serious injury and inconvenience because of this smog. Assuming that damages rather than an injunction are ultimately requested, they must also demonstrate that their damages can be reasonably evaluated in terms of dollars.

The defendants also must convince the court of their theory of "defective condition," which includes not just the specific defects of particular engines, but *all* engines of *all* motor vehicles currently in use in Los Angeles which are manufactured by the defendant motor car companies, and which emit exhaust which is a major component of Los Angeles' smog. The suit does not argue that the cars perform other than precisely as the manufacturers intended, and as their buyers expected. Most "defective condition" cases decided by the courts involve defective physical mechanisms, for example the failure of an altimeter to register the correct altitude, which leads to the crash of an airplane. The only cases analogous to the smog case have concerned food or drugs, which involved some risk or harm, and these have been ruled to be defective only when they are both "unreasonably dangerous," and when the manufacturer had reason to know of the danger but provided no warning.[23]

The case might turn on whether the plaintiffs can introduce evidence that contemporary motor vehicle technology is capable of eliminating major pollutants from automobile exhausts, or alternatively that different forms of carburetion exist that could utilize non-polluting forms of gasoline. The defendants would also have to indicate that automobiles embodying the improvements indicated could be priced and sold in numbers sufficient to maintain an economically viable automobile industry. At this point, the plaintiffs could argue that it was in the public interest to internalize the external costs of smog in the price of the automobile via the incremental costs of pollution control equipment.

If on the basis of such evidence the court were to support a finding of a smog-producing exhaust as being a "defective condition" and unreasonably dangerous, the court could consider remedies against the defendants, presumably with a balancing of equities, which would rule out either a $15 billion settlement, or a blanket injunction against internal combustion engines which emitted smog-producing exhaust. The key factor, from the standpoint of environmental law, would be the definition of defective condition, which would open the door to the use of a products liability theory as a much more effective weapon in pollution cases.

606. Abnormally Dangerous Activities

An abnormally dangerous (or extrahazardous, or ultrahazardous) activity is one which necessarily involves a risk of serious harm to persons or goods, and which cannot be eliminated simply by the exercise of extreme care. In *Luthringer* v. *Moore*,[24] an exterminator was held liable for damages resulting from hydrocyanic acid gas leaking from premises in which he used it to kill cockroaches. The precautions he had taken were considered appropriate, but the activity itself was ruled as being ultrahazardous by nature, and the defendant was thus liable for any damages which might occur, independent of the existence of any fault.

There are at least two major applications of the theory of abnormally dangerous activities to environmental law. A number of states have held the activity of drilling an oil well to be an ultrahazardous one which subjects the owners and drillers to liability if accidents occur. On February 20, 1969 the State of California, County of Santa Barbara, and Cities of Santa Barbara and Carpinteria filed suit against Union Oil, Mobil Oil, Gulf Oil, Texaco, and Peter Bawden Drilling, Inc. for injuries resulting from the blow-out of a well being drilled on part of the continental shelf in the Santa Barbara channel. The claim was for "no less than $500,000,000," and the damage was that petroleum "was deposited into and onto the waters, lands, fish, wildlife, and personal property of the State and all plaintiffs." As yet there is no ruling on whether liability without fault applies in this case.[25]

The second major application of the abnormally dangerous activities theory concerns environmental damage resulting from noise levels, sonic booms, and upper-atmospheric pollution resulting from flights by supersonic transport planes (SSTs). All editions and drafts of the *Restatement of Torts*, an outline of law usually followed by the courts, have regarded the operation of an SST as an abnormally dangerous activity which produces strict liability for ground injuries to persons or property caused by noise or sonic booms.[26] The issue of damage to the stratosphere from engine exhaust has not been treated either in legal cases or in the *Restatement*, but it is likely that the treatment would be the same.

Other than the applications of oil-well drilling and SSTs, a plaintiff in a pollution case would probably find it difficult to successfully argue abnormally dangerous activities unless he could also argue nuisance and/or trespass. To date the successful cases have been so few as to have had a negligible effect on internalizing the social cost of polluting activities.

607. Riparian Rights

The riparian doctrine concerns the right of each user of land on a waterway to a coequal use of water in the waterway. Persons whose waterway has been polluted may file a lawsuit against the alleged polluter.[27] Most states use the so-called "reasonable use" concept of riparian rights, which means that waste disposal may be reasonable in relation to the rights of other riparians (those who abut on the waterway). A definition of "reasonable" requires a balancing of equities similar to that carried out in nuisance or trespass law.

However, courts in determining "reasonableness," have tended to consider only economic aspects, and have rejected arguments as to the natural beauty or public health factors of the water. In the 1965 *Kennedy* case,[28] the court held that: "A riparian owner has no proprietary right in a beautiful scene presented by a river any more than any other owner of land could claim a right to a beautiful landscape." Such interpretations of riparian rights are probably of little value in environmental protection, except perhaps in highly specialized cases.

608. Public Trust Doctrine

A new and promising, but as yet untested approach to environmental protection law which relies on the so-called public trust doctrine has been proposed by Joseph Sax.[29] Public trust law considers that rivers, seashores, and public land are held in trust for the benefit of the public. Sax claims that "the function which the courts must perform . . . is to promote equality of political power for a disorganized and diffuse majority by remanding appropriate cases to the legislature after public opinion has been aroused." The function of the courts in the public trust area therefore will be one of insuring that the democratic process is followed. In a dispute between hunters and fishermen, and those who wished to drain a salt marsh to build luxury condominiums, the court would refer the case to a local, state, or federal political entity and then, having made sure that one interest is not underrepresented in the political process, the court would withdraw.

An example of a public trust case is *Gould* v. *Greylack Reservation Commission*,[30] a 1966 case in which a Massachusetts public parks commission had leased 4,000 acres of public land to a private organization for the development of a ski area. Five citizens of the county in which the land was located brought suit as beneficiaries of a public trust under which the land was held. The Supreme Judicial Court of Massachusetts held the lease invalid, questioning why the state would subordinate a public park to the use of private investors for a commercial development. Under the public trust doctrine as proposed by Sax, the plaintiffs might ask the court to invalidate the lease until the Massachusetts Legislature could vote on whether the interests of the citizens of the state were being served. Sax suggests several guidelines that a court might use in determining whether a particular resource decision has been improperly handled at the administrative or legislative level. Perhaps the most important of these guidelines are questions of whether the public property has been disposed of at less than market value where there is no obvious reason for the grant of a private subsidy, and whether the resource is being used for its natural purpose. The underlying assumption here is that a natural resource like a forest has its most beneficial public use when left in its natural condition.

609. Writ of Mandamus

A writ of mandamus is a court order directed to a public official which requires the performance of the public duties of his official office.[31] A number of studies of environmental protection problems have found that chronic nonenforcement of existing regulations, rather than the lack of effective regulations, has been the central characteristic of the failure of administrative agencies to protect the environment.[32] If an administrator's action involves an exercise of administrative discretion on a given question, then the courts will normally not interfere with the judgment because it is discretionary. However, where an interpretation is clear from the authorizing statute and the administrator or agency is acting erroneously, then the court can compel the agency to do its duty.

Mandamus achieved an important role in environmental protection with the passage of the National Environmental Policy Act in 1969, which requires all federal legislation and all actions of federal administrative agencies to include a statement on the impact of the proposal on man's environment, any adverse environmental effects which are unavoidable, alternatives to the proposed action, and the relationship between short-term uses of the environment and the maintenance and enhancement of long-term productivity. Thus, a full consideration of environmental factors by governmental agencies is now a congressional mandate, and a private citizen concerned about the effects of an agency's activity can seek a judicial review of the action by charging noncompliance with the National Environmental Policy Act. One suit brought under mandamus caused postponement of construction of the Trans-Alaska pipeline pending review of the environmental implications of the line, and consideration of alternative ways of moving oil from the Prudhoe Bay area to United States markets.

610. Stockholder Suits

At present, Security and Exchange Commission rules do not require management to submit to stockholders, proposals "primarily for the purpose of promoting general economic, political, racial, religious, social or similar causes," which probably includes environmental protection. However, if a stockholder can point to some specific economic detriment which might result from a corporation's continuing pollution, such as a possible damage judgment, then he could file a suit. The suit would ordinarily be based upon the theory of waste of corporate assets through management's failure to purchase and install pollution control equipment, so long as the expected cost of damage judgments in favor of victims of pollution exceeded the cost of the necessary control equipment.

Regardless of any waste of corporate assets, a stockholder might bring suit against pollution which exceeded government established emission levels on the basis that the corporate officers were guilty of a *per se* breach of fiduciary duty by being in violation of the law.

While there is some evidence that increased stockholder pressure is being brought to bear on polluters, it is problematic whether sufficient numbers of environmentally concerned stockholders exist to make the stockholder suit a more viable means of harassing reluctant managements.

A variant of the stockholder suit is exemplified by the General Motors proxy fight ("Campaign GM") in the spring of 1970. In late 1969, a group of Washington attorneys, working through the Project on Corporate Responsibility, drafted and submitted to General Motors management nine proposals dealing with minority employment, warranties, air pollution, the composition of the GM Board of Directors, and the formation of a shareholders' committee to advise management. GM management refused to place any of these proposals on the 1970 proxy ballot, and the Project lawyers appealed to the Securities and Exchange Commission. The SEC ordered, without any explanatory opinion, that two proposals be included on the proxy—one dealing with the formation of a shareholder committee to "prepare a report and make recommendations to the shareholders concerning the role of the corporation in society," the other to amend the by-laws to expand the Board to include three new "public interest" directors. At the May 22, 1970 stockholders meeting, the shareholder committee and expanded board proposals received the votes of 2.73 percent and 2.44 percent of the total shares and 7.19 percent and 6.22 percent of total shareholders respectively. The proxy proposals received support from eleven universities and colleges, eight religious organizations, and seven other organizations including the pension funds of New York City and San Francisco. The Project has indicated that it will fight future proxy battles with GM in the same manner, but that it will place major emphasis on the decision-making process and on possibilities for structural reform.

611. Constitutional Arguments

There may be a constitutional argument for the right to a pollution-free environment under the guarantees of the Fifth, Ninth, and Fourteenth Amendments to the Constitution. According to this argument, environmental degradation leads to a deprivation of "life, liberty, or property, without due process of law" [Amendments Five and Fourteen], and although the Constitution is not explicit about the right to be free from pollution, this is overcome by the Ninth Amendment, which reads: "The enumeration . . . of certain rights, shall not be construed to deny or disparage others retained by the people." [33]

In the *Hoerner Waldorf* case,[34] the Environmental Defense Fund sought an injunction on the constitutional argument that continued emission of sulphur oxides by Hoerner violated the rights of citizens guaranteed under the Ninth Amendment, and also violated the due process and equal protection clauses of the Fifth and Fourteenth Amendments. The argument used in the case is as summarized above, with the addition that it cites the Warren Court's decision in a right-to-birth-control-information case[35] that each of the specific rights listed in the Bill of Rights has "penumbras . . . that help give them life and substance." It is reasoned from this that there must surely be a right to an environment free from environmental poisons. A victory by the Environmental Defense Fund in the *Hoerner Waldorf* case would stand alongside *Brown v. Board of Education*[36] in its significance to the future of American constitutional law, and in its impact on society.

Several states have already introduced amendments to their own constitutions to formalize the right to a non-polluted environment. Article One of the Constitution of Pennsylvania was recently amended to read, in part:

> The people have a right to clean air, pure water, and to the preservation of the natural scenic, historic, and esthetic values of the environment. Pennsylvania's natural resources . . . are the common property of all the people, including generations yet to come. As trustee of these resources, the Commonwealth shall preserve and maintain them for the benefit of all the people.

612. Refuse Act of 1899

The Refuse Act of 1899 (more formally, the Rivers and Harbors Act of 1899), is a recently "rediscovered," and potentially very powerful tool in the hands of those concerned with water pollution control. Section 13 of the Act prohibits anyone, including individuals, corporations, municipalities, or other governments from discharging any "refuse" into navigable lakes, rivers, streams, or into the tributaries of such waters. The term "refuse" has been defined by the Supreme Court to include all foreign substances and pollutants, whether accidentally discharged or not, and whether valuable or of no value. The only exceptions to the Act are liquid sewage, and other materials if a permit is obtained from the Army Corps of Engineers. Although the Act enjoins United States attorneys "to vigorously prosecute all offenders," it has rarely been used until recently—largely because the Army Corps of Engineers has not insisted on permits, and has not bothered to bring offenders to the attention of the attorneys.

The Act provides criminal penalties of fines from $500 to $2,500 a day, or imprisonment from 30 days to a year. More importantly, it permits injunctions against continued dumping. Public service lawyers representing organizations such as the Environmental Defense Fund have seized upon this resurrected statute because it also provides that citizens bringing to United States attorneys information leading to conviction are entitled to half the fine. If the attorney does not act on the information supplied, the complaining citizen, in a *qui tam* action, can file suit himself.[37] The fine is important to environmental groups for the incentive it offers industrial employees and others to bring information to the attention of United States attorneys and to file suit if no action is taken.

The first person to be prosecuted under the criminal provisions of the Act was J. J. O'Donnell, President of the J. J. O'Donnell Woolens Company which had for decades poured dyes and soapy waste water into the Blackstone River in Grafton, Massachusetts. The conviction of Mr. O'Donnell in November of 1971 signalled a new approach of holding corporate executives criminally responsible for the water pollution caused by their plants. The theory behind this penalty is that recalcitrance will be overcome more quickly by this method than by the old practice of bringing criminal and civil charges against a company—which can hardly be thrown in jail.

The use of criminal law provisions against individuals such as Mr. O'Donnell raises a number of delicate moral, economic, and political questions. In a society in which corporations are run on the profit motive and pollution is a fact of industrial life, is it fair to penalize individuals or is it simply a hunt for a scapegoat? Is criminal action against a handful of individuals an effective mechanism for reducing pollution, or is it merely a political palliative?

Legally, there is no doubt about corporate responsibility. Executives have long been held responsible for the actions of their companies in antitrust and fraud cases. But pollution raises different kinds of issues. For one, criminal prosecutions tend to promote the "demonology myth" that pollution is caused by a few greedy and callous industrialists. The O'Donnell case bears witness to this argument. Mr. O'Donnell responded to his criminal conviction (and relatively low $2,500 fine) by terminating operations and by moving his plant out of Massachusetts. His textile mill, in what was once a thriving mill town, discharged its 160 employees.

Although the trend is to prosecute individuals, corporations are also beginning to be indicted under the Act, sometimes with remarkably large fines. The most dramatic example occurred in 1971 when Anaconda, which for years had dumped copper fragments from its wire-making plant at Hastings-on-Hudson in New York into the Hudson River, was fined $200,000 under the Act on a hundred-count indictment covering the first five months of 1971. The fine was by far the largest under the Act and is many times larger than the penalties so far levied under state antipollution legislation. Judge Thomas Croake commented that "pollution levies can no longer be shrugged off by corporations as a cost of doing business," and Anaconda would appear to agree—within five weeks it had completed installation of settling tanks to remove the copper from its discharges into the Hudson.

The likelihood that the Act will receive a broad interpretation is greatly increased by the recent flow of executive orders designed to minimize pollution, maximize recreation, and preserve natural resources. If the Refuse Act remains in force, and if the courts allow plaintiffs to file private *qui tam* actions and to collect penalty fees, the Refuse Act could become the most powerful single legal tool for preventing the pollution of navigable waters in the United States.

613. Other Legislation

It is obvious that there are situations in which immediate and decisive regulatory action is the only sensible approach to environmental pollution. Even if effluent fees or taxes are generally preferable to specific regulations, it is unlikely that we would ever regret simply having forbidden the disposal of heavy toxic metals like mercury where they can be consumed by animals or humans. We may also want to stop irreversible deterioration of certain natural resources at once. These are cases where the necessary user charges or taxes would be prohibitive.

This section will briefly review the provisions of federal pollution control statutes other than the Refuse Act. In general, little judicial history of these statutes is available to aid in assessing their value in bringing about pollution control or cost internalization. The most important statute, the National Environmental Policy Act of 1969, will be discussed in detail in Chapter Seven.

The Federal Water Pollution Control Act is the most important federal legislation on water pollution.[38] Under the Act, the pollution of interstate or navigable waters which endangers the health or welfare of any person is subject to abatement. Interstate waters are defined as all rivers, lakes, or other waters that flow across or form a part of state boundaries, including coastal waters such as the Great Lakes. Section Eleven of this Act provides for liability without fault for damages to any publicly-owned or privately-owned property resulting from the discharge of any oil, or from the removal of any such oil. Owners and operators of refineries and off-shore drilling rigs are liable to the government for the cost of removing spilled oil from the waters and shoreline up to a maximum liability of $14 million, and may also be fined up to $10,000. Curiously, the Act establishes strict liability to the federal government, which is capable of defraying cleanup costs from its own funds, but specifically denies strict liability to private actions in which the plaintiffs cannot afford to absorb their losses.

The Oil Pollution Act of 1961 makes it unlawful for a tanker or other ship to discharge oil within fifty miles from shore. Violation of the Act is a misdemeanor and carries a fine of from $500 to $2,500, or imprisonment not to exceed one year for each offense. There appear to have been no charges ever laid under this Act.

The Air Quality Act of 1967[39] was intended as a blueprint for an effort to deal with air pollution problems on a regional basis. The Act grants authority to the Secretary of Health, Education and Welfare to seek injunctions to abate the emission of contaminants anywhere in the country; to design "air quality control regions" for the purpose of implementing air quality standards; in the absence of effective state action, to establish and enforce ambient air quality standards for each region; and to establish federal interstate air quality planning commissions. While no provision is made for private suits or for monetary penalties, a writ of mandamus could certainly be directed at the Secretary of HEW, or at state officials to require performance under the Act.

The Federal Insecticide, Fungicide and Rodenticide Act[40] administered by the Department of Agriculture makes it unlawful to ship in interstate commerce any pesticide (dieldren, DDT, etc.) which is not registered under the Act. The principal purpose of the Act is to prevent misbranded articles from being sold in interstate commerce. However it may also be interpreted as preventing poisons which cannot be used safely from being registered. In the case of *Environmental Defense Fund* v. *Hardin*,[41] petitioners requested an order that the Secretary of Agriculture suspend and then cancel the registrations of pesticides containing DDT under the Insecticide Act. The court ordered the Department to begin cancellation proceedings within 30 days or to give detailed reasons for refusing to initiate proceedings.

There is no general state or federal law to protect the aesthetic quality of the environment, although the National Environmental Policy Act may afford some general protection through subsequent judicial interpretation. However, statutes are passed from time to time which are directed to the aesthetic or visual aspect of some project, area, or type of project. The Billboard laws are an example: The Federal Highway Act authorizes some control over billboards, junkyards, landscaping, and other aesthetic considerations.[42] The federal act does not prohibit a state statute which regulates (not "takes") without compensation or which is stricter than federal requirements.

At the federal level aesthetic considerations are proper under the welfare clause. In *Berman* v. *Parker*, Justice Douglas said:

> The concept of the public welfare . . . represents spiritual as well as physical, aesthetic, and monetary values. It is within the power of the legislature to determine that the community should be beautiful as well as healthy, spacious as well as clean, well-balanced as well as carefully patrolled . . . .[43]

Other such legislation includes a bill in California to ban gasoline-powered automobiles by 1975, and a New Jersey proposal to fine each commercial jet aircraft $2,500 for uncontrolled emissions on each take-off and landing. In the last two years alone, some 550 bills and amendments have been introduced in Congress dealing with the environment—almost certainly the highest concentration of Congressional attention on a single issue since World War Two.

CLASS SUITS

614. Class Actions

A class action is not a cause of action against polluters, but rather a procedure by which a group of persons involved in an issue such as pollution can sue as representatives of a class (or group) of persons even though the group which is suing does not include every member of the class. There are two principal requirements for the maintenance of a class action: the persons constituting the class must be so numerous that it is impractical to bring them all before the court, and the persons filing the class suit must be reasonably representative of all the members of the class. Class members must have aggregate individual claims of at least $10,000 in order to file suit in a federal court.[44] One practical problem with suits as small as $10,000, however, is that they do not offer sufficient contingent fee potential to induce lawyers to take the case on a contingent fee basis.

The economic value of a class action is that it reduces bargaining costs between the polluter and the victim, and among the victims. It enables individuals to combine their bargaining strength at minimal cost, and it presents courts with an aggregate claim which may shift the balance of equities away from the polluter. By presenting an aggregate damage claim, it may raise costs high enough so that the polluter finds it less expensive to abate his emissions than to run the risk of further class actions. Otherwise, faced with only a few plaintiffs claiming minimal damages, corporate polluters may react as did one executive of Reynolds Metals Company in a pollution suit where he testified: "It is cheaper to pay claims than to control fluorides." [45] A class suit allows a polluter to settle with the class representatives, before or after litigation, knowing that the entire class is represented and will be bound by the negotiated settlement.

The use of class actions in environmental litigation has not been common, although, for example, air pollution by a single firm, affecting a group of victims geographically clustered around the polluter where the harm caused to each individual is similar, would seem to present a classic situation for a class action suit.[46] The environmental class actions that have been filed have produced mixed results. One class action for $39 billion against virtually all industry and all municipal corporations in Los Angeles County was dismissed on the court's own motion.[47] In a more recent case, two Chicago aldermen sued the leading automobile, truck, and tractor manufacturers for $3 billion, alleging that they conspired to delay the research, development, and installation of air pollution control devices on their vehicles.[48]

In the *Storley* case,[49] 56 riparian plaintiffs representing 70 downstream farms brought a class action against the Armour & Company slaughterhouse for polluting a river. The suit was successful, and the court awarded damages. However, *Storley* is a class action only in the sense that one legal action was substituted for a number of separate ones. No firm rules for conditions under which a class action for pollution abatement might be brought, or for defining an acceptable representative of a class of pollution victims, were produced that are applicable to subsequent cases.

Federal agencies such as the National Air Pollution Control Administration have been openly encouraging private and class suits as a flexible means of solving local pollution problems, and of bringing to the forefront local environmental issues which should be solved through legislation; the NAPCA has offered their agency's assistance and technical competence in the conduct of such litigation. Class suits are particularly encouraged by these agencies in cases involving problems such as odors, which are difficult to control through the use of standards.

Given that a clearcut decision in the pollution class action field is still lacking, development of the class action concept into a truly effective cost-internalization tool will remain a challenge to environmental lawyers, and to the ability of the legal system to adapt its procedures and rules to changing technological requirements. Many environmental cases which might be litigated will remain unfeasible unless class plaintiffs can aggregate the total damage resulting from the pollution, and unless multiple polluters can be joined as class defendants, thus eliminating the barrier of proving causation where multiple polluters exist.

LIMITATIONS TO EFFECTIVE LEGAL COST-INTERNALIZATION

615. General Problems

Private suits as a device for implementing environmental protection encounter a number of inherent defects. As indicated earlier, establishing liability confronts sometimes intractable problems of proof, especially when the effects in question are intangible and thus difficult to measure, or where they are interacting and cumulative and thus difficult to attribute to one source. The most serious consequences of environmental pollution— damage to future generations, or the gradual erosion of the quality of human existence, cannot be readily associated with any single polluter or group of polluters, and could not in any event be translated into dollar damages.

Further obstacles to an effective legal role in internalizing industrial pollution costs may take several forms. The passive nature of the courts and their inability to initiate an investigation or injunctive suit, means that the questions reaching them typically involve only well-established problems and entrenched economic interests. Citizens have difficulty approaching the court until the injury is visible, or construction of potentially polluting facilities is well advanced. If the damage is only potential and not actual, the court may refuse to act. In a case against the Atomic Energy Commission, a court refused to consider whether thermal pollution of the Connecticut River ought to be enjoined or at least compensated, because the existing permit authorized only construction, not the operation of the nuclear power plant.[50] In another case, a court refused to issue an injunction against an environmentally-damaging highway project because substantial money had already been spent on land acquisition and preliminary planning.[51]

The considerable expense of private litigation against a corporate defendant severely restricts the number of suits possible. The extended court battle in the *Scenic Hudson* case[52] is estimated to have cost conservation groups more than half a million dollars and the Consolidated Edison Company of New York almost a million dollars. As a rule of thumb, to mount an effective environmental lawsuit where expert witnesses are required and the defendant is well financed will cost a minimum of $100,000. Costs of this magnitude force organizations such as the Environmental Defense Fund to make a careful selection of those suits with the greatest potential for publicity and precedent value.

The lack of predictability of litigation against any specific polluting industry has the effect of reducing the deterrent effect of the potential litigation, and also the incentive for abatement. Since episodic litigation is not foreseeable in the same way that government effluent charges are, and is not automatically triggered by a predeterminable action, industry reaction is frequently one of delaying cost internalization, and fighting suits as a deterrent while instituting only minimal antipollution measures.

616. Uneven Distribution of Costs

The unpredictable nature of private actions against polluters may lead to an uneven distribution of costs to particular enterprises, unduly burdening them to the advantage of those equally guilty of pollution, but who manage to avoid litigation. The social costs involved can be internalized with less strain if all the relevant competitors must internalize them, so that all pricing policies will be affected in similar fashion. It is unlikely that selective and random litigation will result in all forms of waste-producing enterprises being placed under the same cost-internalization requirements. Without a uniform distribution of costs, incentive for expenditures on improved pollution abatement is diminished, since each firm will calculate its probability of avoiding litigation, and will come up with an expected-cost-of-litigation-and-damages figure that is likely to be less than the cost of pollution control.

Those who advocate a broadening of common law actions to effectuate cost-internalization have recognized this problem, and have suggested that statutory arrangements to protect enterprises in selected industries from overwhelming liability is necessary. Such "insurance" arrangements would spread the cost burden of suits widely enough among firms to avoid penalizing any single enterprise that happened to be the target of a lawsuit for an accident that could have happened to any firm within an industry— for example, in offshore oil drilling. However, the risk must not be so widely diffused as to nullify the incentive for all firms to internalize the costs of their operations—this might be accomplished by having a high insurance "deductible," which varied inversely with the amount of pollution control equipment used by each operator.

FUNCTION OF LEGAL APPROACHES TO ENVIRONMENTAL PROTECTION

617. In General

The procedural and legal limitations of achieving effective internalization of pollution costs do not mean that the courts have no useful role to play in environmental protection. Private suits may be used in cases involving types or degrees of pollution that simply are not covered by existing statutes. For example, the 1967 Air Quality Act authorized the Secretary of Health, Education, and Welfare to establish air quality regions and ambient air standards, but specific regulations were not issued until mid-1971, with comprehensive standards still undetermined.

Private suits may also be used to fill gaps in existing enforcement of regulations. As pollution problems become more severe, overburdened agencies will have to become increasingly selective in prosecuting violations of pollution statutes. Private suits can be used to attack violations that agencies must ignore because of manpower and financial constraints. Where specific pollution is damaging to people located near its source but is minor relative to larger-scale pollution in the area, private enforcement through litigation may be the only way to force abatement.

Private suits may also provide a more sophisticated tool than regulatory response in dealing with the rapidly changing technology of pollution abatement. The ability of courts to hear new evidence and to shape novel remedies has led some writers to the conclusion that courts are more responsive than legislatures or administrative agencies to the impact of pollution, and the control techniques that might be applied.[53]

Private suits are certainly flexible from a time standpoint. The enactment of useful legislative change always takes time, a commodity which may become critical when particulate or other pollution suddenly reaches critical levels. An informed citizen can bring the problem to the attention of the court almost immediately; in the short run, the court is the arena in which industry lobbyists are least able to exercise their delaying tactics.

Lawsuits may also lead to the development of new public attitudes towards pollution by serving as a focal point for the gathering and dissemination of assumptions about the application and handling of technology. The very language used in the law may induce cultural and moral change. To require an industry to "internalize the social costs of its pollution" means little to the public at large, but to hold the same polluter liable for damages for the "nuisance" he has caused, or to hold him legally responsible for the "defect" in his products says a great deal, and is widely noted.

618. Publicity Function

An unrecognized benefit of the private suit against a polluter is its potential as a catalyst for inducing change elsewhere in the system. This is true particularly in our litigious society where the media pay front-page attention to the more dramatic court battles. Just as much of the most important civil rights legislation in the United States was induced by newsworthy, although ineffectual litigation, so much of the required legislation to protect the environment might be induced by successful (or even more by unsuccessful) private suits seeking to internalize the costs of pollution. A polluter-defendant in a suit, even if he "wins" the litigation, may be spurred to corrective abatement to avoid further unfavorable publicity.

619. Public Participation

Private suits enable the individual litigant to feel that he is capable of some involvement in the complex industrial activities that affect the quality of his life, that he is more than a passive inhabitant of an environment the future of which is beyond his control. To be accorded a court hearing to help determine whether an activity that harms him will be allowed to continue, and if so on what terms, has value in itself as an affirmation of the individual's right not to be reduced to a means towards someone else's technological or industrial end. The limited number of suits that can realistically be brought, given their expense, can provide direct benefits of participation to only a few. A less direct psychological satisfaction may accrue to the hundreds of thousands of citizens who are currently contributing to the Environmental Defense Fund and similar public organizations to assist their efforts in private litigation on behalf of the environment.

In fairness, something must also be said about the cost and nuisance that can be created for the legitimate corporation by crank suits instituted in spite of the expense involved. The small industrial corporation in particular is vulnerable to suits that are either unfair or vindictive, and which, even when won, can be extremely costly in terms of both money and executive time.

REFERENCES

[1]Victor Yannacone, *Sue the Bastards*, a speech delivered on "Earth Day," April 22, 1970, at Michigan State University, East Lansing, Michigan.

[2]The structure and part of the content of this chapter was inspired by a paper entitled *The Courts and Industrial Pollution Abatement*, by Barry R. Furrow, my research assistant at the Harvard Law School during 1970-71.

[3]For example *Environmental Defense Fund* v. *Hoerner Waldorf Corp.*, Civil No. 1694 (D. Mont., filed November 13, 1968); *Sierra Club* v. *Hickel*, Civil No. 51,464 (N.D. Calif., filed June 4, 1969); *Citizens Committee for Hudson Valley* v. *Volpe*, 302 F. Supp. 1083 (S.D. New York, filed June 28, 1969). An interesting case is *Diamond* v. *General Motors*, No. 947,429, where a multi-billion dollar suit was brought by a citizens group against thirteen major corporations, the Department of HEW, and the United States Attorney General for their alleged role in creating Los Angeles smog.

[4]*Gerring* v. *Gerber*, 219 NYS 2d 558 (1961).

[5]See "Developments in the Law—Injunctions," *Harvard Law Review*, No. 78 (1965), p. 994.

[6]See Coleman, "Possible Repercussions of the NEPA of 1969 on the Private Law Governing Pollution Abatement Suits," *Natural Resources Lawyer* (1970), p. 647.

[7]In an 1896 case, plaintiffs sued because defendant's coal mining operations had fouled the plaintiff's watercourse. The court concluded: "To encourage the development of the great natural resources of a country, trifling inconveniences to particular persons must sometimes give way to the necessities of a great community." Reference: *Penna. Coal Co.* v. *Sanderson*, 113 Pa. 126 (1896), p. 6 A. 459.

[8]*Waschack* v. *Moffat*, 109 A. 2d 310 (1954), p. 316.

[9]For a discussion, see W. Prosser, *Torts*, No. 89 (3rd edition, 1964), p. 488.

[10]*William Aldred's Case*, 77 Eng. Rep. 816 (K.B. 1611), cited in Julian Conrad Juergensmayer, "Control of Air Pollution Through The Assertion of Private Rights," *Duke Law Journal* (1967), p. 1125.

[11]*Madison* v. *Ducktown Sulphur, Copper & Iron Company*, 113 Tenn. 331 (1904).

[12]*Hulbert* v. *California Portland Cement Company*, 161 Cal. 239 (1911).

[13]*Renken* v. *Harvey Aluminum, Inc.*, 226 F. Supp. 169 (D., Oregon 1963).

[14]Milton Katz, *The Function of Tort Liability in Technology Assessment* (Cambridge, Mass.: Harvard University Program on Technology and Society) Reprint Number 9 (1969), p. 615.

[15]*Jost* v. *Dairyland Power Company*, 172 N.W. 2d 647 (Wisc.,1970).

[16]*Boomer* v. *Atlantic Cement Company*, 309 NYS 2d 312 (1970).

[17]See Patrick E. Murphy, "Environmental Law: New Legal Concepts in the Antipollution Fight," *Missouri Law Review*, No. 36 (1971), pp. 81-82. For an expansion of the "Taking of Property" concept see R. Lester, "Nuisance As A 'Taking of Property'," *University of Miami Law Review*, No. 17 (1963) p. 537.

[18]*Martin* v. *Reynolds Metals Company*, 342 P. 2d 790 (1959), *cert. denied* 362 U.S. 918 (1960).

[19]Cases similar to *Martin* in that they also involved the Reynolds Aluminum plant at Troutdale, Oregon but which had quite different outcomes are *Arvidson* v. *Reynolds Metals Company*, 125 F. Supp. 481 (W. D.,Wash. 1954), *aff'd.* 236 F. 2d 224 (9th Cir. 1956), and *Fairview Farms* v. *Reynolds Metals Company*, 176 F. Supp. 178 (D. Ore. 1959), both of which are discussed in Juergensmeyer, "Control of Air Pollution Through the Assertion of Private Rights," *op. cit.*, pp. 1138-1142. The history of applying a trespass theory to water pollution cases is limited. Courts have allowed recovery under a trespass theory for damage to oyster beds resulting from dredging operations causing silt, or from sewage discharged into oyster beds. See *Mason* v. *United States*, 123 Ct. Cl. 647 (1952).

[20]*Greyhound Corporation* v. *Blakely*, 262 F. 2d 401 (9th Cir. 1958).

[21]*Martin* v. *Reynolds Metals Company*, 342 P. 2d 790 (1959).

[22]C. Jon Handy and William R. Bernstein, *For Themselves and On Behalf of The People of The United States That Are Similarly Situated* v. *General Motors, Inc.*, Civil Action No. 69-1548-R (C.D. Cal., August 7, 1969). The short title and the discussion of the *Handy* case come from Milton Katz, *op. cit.*, pp. 623-638.

[23]There are a vast number of drug products liability cases to choose from. A recent one is *Basko* v. *Sterling Drug, Inc.*, 416 F. 2d 417 (1st Cir. 1969).

[24]*Luthringer* v. *Moore*, 190 P. 2d 1 (1948).

[25]*State of California* v. *Union Oil Company*, No. 84594, Superior Court of Santa Barbara County, California (February 20, 1969). Since the Santa Barbara blow-out, federal law [section 250.42(a)] has been amended to make pollution of the high seas by drilling or production of

oil by definition an abnormally dangerous activity, thus imposing the rule of liability without fault on such activities. To not impose liability in such cases would be equivalent to imposing a tax on victims of blow-outs and turning the proceeds over to the oil industry as a subsidy.

[26]See *Restatement (Second) of Torts* 520A, and comments a through d (1964).

[27]The riparian doctrine is accepted in 31 eastern states. In most of the western states the rule of prior appropriation ("he who gets there first, controls") is followed.

[28]*Kennedy* v. *Moog Servocontrols, Inc.*, 264 N.Y.S. 2d 606 (Sup. Ct. 1965).

[29]Joseph Sax, "The Public Trust Doctrine in Natural Resources Law: Effective Judicial Intervention," *Michigan Law Review*, No. 68 (1970), at 473.

[30]*Gould* v. *Greylack Reservation Commission*, 215 N.E. 2d 114 (1966).

[31]Strictly speaking, Rule 81(b) of the Federal Rules of Civil Procedure abolished mandamus, but it preserved the "relief heretofore available by mandamus." The courts have generally held that the remedies available before adoption of the new Federal Rules of Civil Procedure are still available under the new rules, and under the same principles as formerly governed its enforcements.

[32]See E. F. Cox, R. Fellmeth, J. Schultz, *The 'Nader' Report on the Federal Trade Commission* (New York: Grossman Publishers, 1969).

[33]See John C. Esposito, "Air and Water Pollution: What To Do While Waiting For Washington", *Harvard Civil Rights—Civil Liberties Law Review*, No. 32 (1970), pp. 45-51.

[34]*Environmental Defense Fund* v. *Hoerner Waldorf Corp.*, Civil No. 1694 (filed D. Mont., November 13, 1968).

[35]*Griswold* v. *Connecticut*, 381 U.S. 479 (1965).

[36]*Brown* v. *Board of Education*, 347 U.S. 483 (1954).

[37]A *qui tam* action is "brought by an informer, under a statute which establishes a penalty for the commission or omission of a certain act, and provides that the same shall be recoverable in a civil action, part of the penalty to go to any person who will bring such action and the remainder to the state or some other institution. . . ." In effect, the plaintiff sues for the mutual benefit of the state and himself. There is some question as to whether Congress intended citizens the right to bring *qui tam* actions to enforce the Refuse Act. The Act does not explicitly state that citizens have a right to sue directly, nor that they

do not. The prevailing view seems to be that citizen suits are possible. For a contrary argument, see "Commentary: Oil and Oysters Don't Mix—Private Remedies for Pollution Damage to Shellfish," *Alabama Law Review*, No. 23 (1970), pp. 121-124.

[38]33 U.S.C. 466, *et seq*; amended April 3, 1970 by the Water Quality Improvement Act of 1970, Public Law 91-224.

[39]81 Stat. 485, 42 U.S.C. 1857 *et seq*.

[40]7 U.S.C. 135 *et seq*.

[41]*Environmental Defense Fund* v. *Hardin* (D.C. Cir. May 28, 1970).

[42]23 U.S.C. 131 and 136.

[43]*Berman* v. *Parker*, 348 U.S. 26 (1954), p. 33. For a similar statement see *State* v. *Wieland*, 69 N.W. 2d 217 (Wisconsin 1955).

[44]One problem arises from the recent *Snyder* case, which held that class members in a federal diversity case must each meet the jurisdictional requirements in order to aggregate their claims to the required level. *Snyder* v. *Harris*, 394 U.S. 332 (1969), interpreting 28 U.S.C. 1331(a) (1964).

[45]Cited in John C. Esposito, "Air and Water Pollution: What To Do While Waiting For Washington," *op. cit.*, p. 36.

[46]Actions under the relevant statute, Rule 23(b)(3) of the Federal Rules, have arisen almost exclusively in antitrust and securities law situations.

[47]*Diamond* v. *General Motors, et al.*, No. 947,429, California Superior Court.

[48]*Kean, Wigoda et al.* v. *General Motors, Ford Motor Company, Chrysler Corporation, et al.*, No. 69c-1900, U.S. District Court for the Northern District of Illinois.

[49]*Storley* v. *Armour & Company*, 107 F. 2d 499 (8th Cir., 1939).

[50]*New Hampshire* v. *Atomic Energy Commission*, 406 F. 2d 170 (1st Cir. 1969).

[51]*Town of Bedford* v. *Boyd*, 270 F. Supp. 650 (S.D. New York, 1967).

[52]*Scenic Hudson Preservation Conference* v. *Federal Power Commission*, 354 F. 2d 608 (2nd Cir. 1965).

[53]There is considerable difference of opinion in the literature on this point. For a discussion see R. Stepp and S. Macaulay, "The Pollution Problem," in *Legislation and Special Analyses of the American Enterprise Institute for Public Policy Research*, 90th Congress, 2nd Sess. (No. 16, 1968), p. 12.

7

The National

Environmental

Policy Act

"A nation's history is written in
the book of its words, the book of
its deeds, the book of its art. A
people's history is also written
in what they do with the natural
beauty Providence bestowed upon
them."

Richard M. Nixon[1]

700. Introduction

Although Congress has been concerned with environmental legislation since the 1950's, the most comprehensive legislation enacted to date is the National Environmental Policy Act of 1969 (NEPA), signed into law by President Nixon on January 1st, 1970.[2] The statute marks an important departure from existing federal environmental legislation in that it recognizes in Title I the direct interest of the federal government in working toward a healthful environment, rather than placing primary responsibility for environmental legislation on the states.[3] Title II of the Act sets up a Council on Environmental Quality (CEQ) in the Executive Office of the President. The functions of the Council are to review and appraise the various programs and activities of the federal government in light of the policy set forth in NEPA. Thus, the main thrust of the Act is to insure that federal departments and agencies proceed in their activities with due concern for environmental quality.

228

Examples of the rising public concern over the way in which federal policies and activities have contributed to environmental decay may be seen in the Santa Barbara oil well blowout; the impact of a jet airport adjacent to the Everglades National Park in Florida; the indiscriminate siting of steam-fired power plants; the dangers of spillage or leakage inherent in the various methods suggested for moving crude oil from the North Slope discovery in Alaska; the loss of publicly owned seashores and open spaces to industry and commercial developers; and federally sponsored or funded construction activities such as highways, airports, and other public works projects which proceed without reference to the desires of local residents.[4] NEPA is designed to deal with the basic causes and long-range implications of these environmental problems.

Section 101 of NEPA declares that it is national environmental policy that the federal government use all practicable means, consistent with other essential considerations of national policy, to improve and coordinate federal plans, functions, programs, and resources so that the nation may:

(1) fulfill the responsibilities of each generation as trustee of the environment for succeeding generations;

(2) assure for all Americans safe, healthful, productive, and aesthetically and culturally pleasing surroundings;

(3) attain the widest range of beneficial use of the environment without degradation, risk to health or safety, or other undesirable and unintended consequences;

(4) preserve important historic, cultural, and natural aspects of our national heritage, and maintain, wherever possible, an environment which supports diversity and variety of individual choice;

(5) achieve a balance between population and resource use which will permit high standards of living and a wide sharing of life's amenities; and

(6) enhance the quality of renewable resources and approach the maximum attainable recycling of depletable resources.

Under the Act, federal officials and agencies must consider whether adverse environmental impact will result before making *any* decision which affects the environment. Such consideration must be included in a "detailed statement" when legislative proposals or other major federal action are involved.

Federal officials must thus analyze the environmental impact of their proposed actions. Failure to consider an important environmental issue or inadequate consideration of such an issue is grounds to find noncompliance with NEPA where, had the issue been adequately considered, the decision might have been different. In legal terms, if a federal administrator does not adequately consider the environmental impact of his proposed actions, his decision to undertake the actions is considered arbitrary and capricious and subject to reversal in the courts. Further, a federal agency cannot justify a nonconsideration of environmental effects on the basis of ignorance of such effects. If there is inadequate knowledge about a particular problem, NEPA provides authority for the agency to obtain the required information.

If adequate consideration of environmental impact indicates that an action will have beneficial or neutral environmental effects, then the action is permissible under NEPA. If adverse environmental effects are shown, then the official or agency has a duty to consider alternatives to the proposed action. The consideration of alternatives must be as thorough as the consideration of environmental effects; inadequate consideration of an alternative is sufficient grounds to find noncompliance with NEPA, again because, had the alternative been adequately considered, it might have been adopted rather than the original proposal.

If an alternative is found which does not entail the adverse environmental effects of the proposed action, then the duty of federal officials under the Act requires that the alternative rather than the original proposal be adopted. If a conflict can be eliminated by the adoption of an environmentally nondestructive course of action, then it would be a breach of duty to adopt the environmentally destructive course of action.

Some proposals which have adverse environmental effects will not have alternatives which eliminate the adverse effects. In these cases, the federal official or agency has a responsibility under NEPA to reassess the justification for the proposed action. Senator Henry M. Jackson of Washington has pointed out that Congress intended that environmentally destructive courses of action be only infrequently permitted:

The basic principle of the policy is that we must strive in all that
we do, to achieve a standard of excellence in man's relationship
to his physical surroundings. If there are to be departures from
this standard of excellence they should be exceptions to the rule
and the policy. And as exceptions, they will have to be justified in
the light of public scrutiny as required [by the Act].[5]

Under NEPA, the only permissible actions which have adverse
environmental consequences are those where the long-term resulting social benefits outweigh the long-term environmental costs.
The rules established by NEPA to govern this balancing of social
equities are as follows:

First, values other than economic values are to be included in
the weighing. Such values as diversity, aesthetics, and health are to
be considered when determining whether an environmentally destructive action is to be permitted. To the extent that these values
cannot be quantified, other procedures must be developed to insure
that they receive "appropriate consideration."

Second, a much more thorough look at "public benefits" must
be taken than is true under the "balancing of equities" test applied
in nuisance cases. The benefits which accrue to the public, rather
than to individuals, must be balanced against the losses accruing
to the public.

Even where an agency succeeds in proving that an environmentally destructive action is justified by offsetting social benefits,
it must take all possible steps to minimize the adverse effects of
its action. In particular, consideration of alternative techniques of
implementing the decision is important.

Throughout the process outlined, the burden of proof falls on
the person or group which wishes to disturb the environment.
This is a critical point, because it restructures a decision-making
process which has virtually always subordinated the public's interest in environmental protection to a multitude of private
interests. Those who wish to disturb the environment must now
prove either that the proposed action will not impair environmental
quality, or that social benefits will clearly outweigh social costs.
They must also prove that no alternatives exist which would
eliminate or minimize such effects, and the proof must be offered
as part of a reviewable record.[6]

701. Procedural Duties

NEPA requires the preparation of a written statement, to be reviewed by the President, the Council on Environmental Quality, and the public, on the environmental impact of proposals for legislation which "significantly affect the quality of the human environment." [7] In the case of proposals for legislation, the written statement is supposed to allow Congress to determine whether a proposal is consistent with the national environmental policy without the need to plow through the administrative record. If the statement were not required, possible adverse effects would often be buried in the administrative record and would go unnoticed. When the statement is submitted, it must be accompanied by the comments of "appropriate Federal, State, and local agencies, which are authorized to develop and enforce environmental standards." The statement is specifically required to be an analysis of environmental impact rather than an attempt to justify a particular decision. The analytical character of the statement is emphasized by the NEPA requirement that the statement must be prepared by a "responsible official."

Reproduced below is an abbreviated copy of the U.S. Department of the Interior's environmental statement for the proposed Narrows Unit of the Missouri River Basin Project in Colorado, which was submitted in June of 1970 to satisfy NEPA requirements. Prior to its submission to the Council on Environmental Quality, the report had been reviewed by the member states of the Missouri River Basin, the Secretary of the Army, and several unnamed "interested Federal agencies." No revision was made based on the recommendations and views received.

It is of some interest that the environmental statement was transmitted to the Council on June 9th, 1970. Hearings on the House version of the bill approving the project (H.R. 6715) had been held the previous April 16th and 17th by the Subcommittee on Irrigation and Reclamation of the Committee on Interior and Insular Affairs, and the measure had been recommended for passage by the House. Hearings on the Senate bill (S. 3547) were scheduled for the next day, June 10th, by the Water and Power Resources Subcommittee of the Senate Committee on Interior and Insular Affairs.

In reading the environmental statement, consider to what extent it covers the five required factors: the environmental impact of the proposed action; whether there are any adverse environmental effects that cannot be avoided; whether there are alternatives to the proposed action; the relationship between long-term and short-term uses of the environment, and long-term maintenance of productivity; and, whether there are irreversible and irretrievable commitments of resources which are involved in the proposed action. Was the report written by an engineer, an economist, or an environmental protection expert?

ENVIRONMENTAL STATEMENT ON PROPOSED NARROWS UNIT, MISSOURI RIVER BASIN PROJECT, COLORADO, SUBMITTED IN CONFORMANCE WITH SECTION 102 (2) (C) OF THE NATIONAL ENVIRONMENTAL POLICY ACT OF 1969*

Nature of Activity

The proposed Narrows Unit of the Missouri River Basin Project is a multiple-purpose water and related land resources development located in the lower South Platte River Basin in northeastern Colorado. The proposed Unit would serve the functions of irrigation, flood control, recreation, and fish and wildlife enhancement, as well as potential future municipal and industrial water supplies. . . .

Description and Purpose of Proposed Development

The principal feature of the Narrows Unit would be Narrows Dam and Reservoir, to be constructed on the South Platte River near Fort Morgan. The Narrows Dam would be an earthfill structure about 146 feet high with a crest length of 22,100 feet. Three dikes having a combined length of about 12,700 feet also would be necessary. . . .

Rights-of-way adequate for construction and operation and

---

* Statement *by the United States Department of the Interior concerning the proposed Narrows Unit of the Missouri River Basin Project (June, 1970).*

maintenance of the dam and reservoir, associated relocations, and for recreation and fish and wildlife developments would require the acquisition of approximately 36,250 acres of land. Relocation of the Union Pacific Railroad and State Highway 144 would be required.

The construction of a fish hatchery and rearing ponds and the acquisition and development of the existing Jackson Lake Reservoir, now privately owned, are proposed for outdoor recreation and fish and wildlife enhancement. A wildlife management area and four public-use recreation areas are also proposed for development.

Because of inadequate water supplies, the areas irrigated within the lower South Platte River Basin, including the Narrows Unit service area, have been limited, and the full irrigation potential has failed to materialize. The frequent lack of sufficient surface water supplies has caused many irrigators to construct wells for pumping ground water for supplemental irrigation. Water supply shortages have been further intensified by severe droughts, which cause serious depletions of the surface water supply and result in a greater demand on the wells.

In the Narrows Unit service area there are 33 irrigation systems; none of these reservoirs are large enough to store adequate supplies for their associated ditch system. . . .

The lack of storage facilities is a major factor contributing toward the annual shortages of water. Twenty-three of the ditches, which serve 98.4 percent of the irrigable lands in the Lower South Platte Water Conservancy District, experienced an average annual diversion shortage of 178,000 acre-feet over the 1947-1961 period.

The water supply for the Unit would average 140,700 acre-feet annually, of which 119,400 acre-feet would be obtained from regulation of surplus streamflow and from direct-flow water rights associated with irrigated lands to be acquired for the Narrows Dam and Reservoir and 21,300 acre-feet from divertible return flows. . . .

Storage water would be released as necessary from the Narrows Reservoir to supplement irrigation within the service area, totaling 166,370 acres of irrigated land in the conservancy district. Supplemental releases would be conveyed downstream in the river channel to the diversion works of existing irrigation systems.

The supplemental water supply for Unit lands, analyzed on the basis of a 100-year period and an interest rate of 3-¼ percent, would yield $1,410,000 direct benefits and $222,000 indirect and public benefits, for total irrigation benefits of $1,632,000 annually.

The South Platte River Basin is subjected to deluge-type rainstorms that are erratic and incredibly violent. During the period 1844 to 1965 nine such storms occurred, resulting in major floods. Numerous smaller, though severe, floods also have occurred. The impact on the area is substantial, causing major losses to property, transportation facilities, irrigation systems, crops, and livestock, with resultant devastating effects on the economy. Operation of Narrows Dam and Reservoir will afford downstream flood protection. The Corps of Engineers estimates flood control benefits will amount to $1,600,000 annually. . . .

Both the construction of Narrows Dam and Reservoir and the rehabilitation of Jackson Lake Reservoir would provide recreation and fish and wildlife benefits. In addition, specific lands and facilities would be required for recreation and fish and wildlife purposes. Recommended minimum downstream flows for fish requirements would be met most of the time by reservoir seepage and normal project operations. Outdoor recreation activities will include picnicking, sightseeing, boating, water skiing, swimming, hiking, and camping. The National Park Service estimates that use will increase from 930,000 visitor days initially to almost 1-1/4 million about 25 years after initial development. The recreation benefits have been evaluated at $1,410,000 annually.

The proposed fish and wildlife measures will jointly serve the purposes of mitigation and local and national enhancement of those resources. Total evaluated fish and wildlife benefits associated with the fishery, hunting, waterfowl use, and wildlife-oriented recreation are estimated to be $552,000 annually.

Adjusted annual equivalent benefits anticipated from development of the Narrows Unit total $5.2 million, of which $5.0 million are direct benefits.

The estimated construction cost of the Unit, based on January 1969 prices, is $68 million. Annual operation, maintenance, and replacement costs are estimated to be $313,000. Annual equivalent Federal costs for a 100-year period of analysis at 3-1/4 percent interest are computed to be $2.75 million.

The ratio of total annual benefits to annual equivalent costs is 1.9 to 1. The ratio of direct benefits to costs is 1.8 to 1.

The total construction costs ($68 million) have been allocated to the functions of the Unit as follows: irrigation, $21.1 million; flood control, $24.4 million; recreation, $15.8 million; fish and wildlife enhancement, $6.6 million; road relocation, $135,000.

Effect of Proposed Development on Quality of the Environment

This assessment of the probable effect of the Narrows Unit on the quality of the human environment reflects the views and recommendations of those Federal and State agencies which participated directly in formulating the recommended plan of development or indirectly through the review process.

(1) Impact on environment. The South Platte River watershed is a broad rolling plain through which the river has formed a wide valley of flood plains and bench lands on river terrains. The bench lands are situated from 20 to 200 feet higher than the flood plains. The river has a gradient of 8 to 10 feet per mile.

The Unit area . . . is normally semi-arid with widely varying annual precipitation. This climate provides abundant sunshine, with warm days and cool nights during the growing season, making the area especially favorable for agriculture and associated industries, although the natural precipitation is adequate to support only highly speculative dryland farming and livestock grazing. The project area does not constitute a unit of environment that is either scarce or unique.

The Narrows Unit will have a favorable impact on the natural environment and economy throughout the South Platte River system within and downstream from the project area. Beneficial impacts would include creation of a new reservoir having a water surface area of approximately 15,000 acres; establishment of a new reservoir fishery; conversion of 5-1/2 miles of low quality warm water stream fishery to a good cold water fishery; the development and operation of 15,765 acres of project lands as a wildlife management unit to mitigate loss and damages to the fish and wildlife resources and to enhance the project for upland game and waterfowl; stabilization of a 2,500-acre offstream reservoir environment specifically to maximize the benefits therefrom for fish and wildlife and recreation purposes. . . . The Unit would provide a much-needed water-oriented recreation outlet for this region of Colorado.

No downstream water quality effects are anticipated which would interfere with present or proposed beneficial uses of water from the South Platte River. Operation of the Narrows Unit will reduce the salinity and sediment content of the downstream flows.

At the present time, the economic environment of the project area is one of instability with wide fluctuations in income from year to year due primarily to variations in rainfall and water supply for irrigation. This has resulted in a deterioration of the well-being of the area residents. The additional water supply, flood protection, fish and wildlife, and recreational developments and opportunities provided by the Unit would contribute substantially to the improvement of the well-being of the residents of the area.

The difference in irrigated crop values produced in the project area has varied as much as $4 million from one year to the next. A study developed for Nebraska indicates that $6.68 of economic activity occurs within the State for one dollar of increased value attributable to irrigated crop production. On this basis, a reduction of $4 million in crop production translates into a total economic activity decrease of about $27 million. Such wide variations have occurred as recently as 1965. Conversely, with the project in place, economic stability occurs and indications are that total economic activity would increase by about $25 million annually.

(2) Adverse environmental effects. Certain adverse effects on the environment will occur with the project. These will include the inundation by the reservoir of 15.5 miles of natural stream environment and approximately 15,000 acres of land. The loss of approximately 1,100 fur bearers and 1,000 ducks annually associated with the habitat will be mitigated as a part of the proposed fish and wildlife development. There will be no loss of unique archaeological or geological features. The project will require the acquisition of a total of approximately 36,000 acres of predominantly agricultural lands and the dislocation of about 40 farm units associated therewith. Three small settlements, involving about 150 families and one cemetery, will require relocation.

(3) Alternatives to the proposed action. There is no alternative course of action to the proposed plan which would create equivalent benefits for meeting social, economic, and environmental objectives within the State of Colorado at comparable economic costs.

(4) Relationship of short-term uses versus long-term needs. The resources which would be committed to this proposed development are not subject to depletion in the same sense as are coal or oil deposits. Water is a renewal resource. Land productivity can be maintained and enhanced with proper care while being irrigated. Therefore, the water and land resources are not lost, nor is their future development for other purposes having higher values precluded.

Thus, a common relationship exists between local short-term use of these resources and the need to maintain and enhance the long-term productivity.

(5) Irreversible commitment of resources. The only resource commitments of an irretrievable nature involved in this proposed development are enumerated under item (2) discussed previously.

702. The Engineering Fallacy

The environmental statement on the proposed Narrows Unit of the Missouri River Basin Project is one of the more straightforward and accurate statements filed under the NEPA requirement in the first 24 months of its existence. It has its obvious problems—one can, and should, question the calculations of benefits which yield a cost-benefit ratio of 1 to 1.9, and a cost-to-direct-benefit ratio of 1 to 1.8, or the use of an agricultural multiplier of 6.68 when such a value is normally never assumed by economists to exceed 1.2 or 1.3. The real problem, though, is one common to virtually every statement filed in the first 24 months, and will be referred to as the engineering fallacy.[8]

A simple example will suffice to illustrate the fallacy. In the New York City area at the present time, there is great pressure and urgency to construct a fourth jetport. Every projection indicates that there will be 85 million passengers using New York airports by 1980, so there must be a fourth, and perhaps a fifth jetport to service them. If you accept the premise in the last sentence, you have already been conned by the engineering fallacy. If there are no additional airport facilities in New York in 1980, there will not be 85 million passengers. If some of the people who really require airport access cannot get it in New York, they will simply locate elsewhere. Some persons will simply not fly, others will use alternate airports at some distance from the city. Some persons will be prepared to accept long delays in "stacked" aircraft, but there are physical and economic limits to how far that "solution" can be extended.

Consider a second example. In 1960, the citizens of California authorized the largest single bond issue ever floated by a single state—billions of dollars for the California Water Plan. The Plan was concerned with transporting the fresh water from the north to the majority of the people, and thus the bond-issue voters, in the south. Every indication was that southern California would require water for an additional 9 million people by 1985. The problem was how to get the water to the people.

The engineering fallacy was apparent again in the California Plan. The problem could easily have been stated (but wasn't) as follows: "How do we get the people to the water?" a problem that would be physically easier, financially cheaper, and ecologically immensely wiser. Going one step further, however, one might ask: Why get the people and the water together at all? State a need, buttress it with growth projection statistics, and the engineering fallacy will focus your thinking on solutions. But who says it is a problem to begin with? If we don't build the dams and aqueducts of the Water Plan, southern California just might *not* grow by 9 million people. People and industries, learning of the forecast water shortage, might go elsewhere, or might stay home. We would not have to pay the cost of the California Water Plan altering virtually every remaining body of fresh water in the state, leaving none in its natural condition. And we would not have to speculate what this might do to the complex ecology of the Central Valley, the land, and the wildlife.

Getting back to the Department of the Interior's environmental statement: point out to an engineer that the South Platte River in northeastern Colorado sometimes floods its lowlands, and he proposes building a dam—or in this case, one dam and three dikes. Point out to him that the dam will eliminate the fish run on the river and he proposes a fish ladder and artificial gravel spawning pits. Point out that one railroad and one state highway will be suddenly under water and he proposes relocating them. Point out that 15,000 acres of natural stream environment and 36,000 acres of good agricultural land, as well as 40 farms and three villages will be inundated, and he proposes calling in a planner to build a model city for the displaced persons. What he will *never* do is reconsider whether the Narrows Unit dam of the Missouri River Basin Project should be built or not. Altogether NEPA may require him to consider the environmental implications of doing nothing, his whole background and training as an engineer distort his perspective.

Since the development of environmentally nondestructive alternatives is an important component of the substantive duty of federal agencies under NEPA, it is important that they do define "alternative courses of action." Nonaction is always one alternative, but consideration of alternatives should not be limited to this pair. However, when considering positive alternatives, the engineering fallacy may arise again in determining which alternatives are economically or technologically feasible, given that feasibility depends on an allocation of resources that NEPA is designed to change.

For example, in deciding whether to build another freeway through downtown Boston, one is tempted to argue that there is no alternative system of mass transportation that is technologically feasible. But it was the original federal decision to subsidize highway construction rather than rapid transit that now makes it technologically and economically unfeasible to build other forms of mass public transportation.

Economic impracticality is often cited as an obstacle to the adoption of environmentally nondestructive alternatives because of an incorrect analysis of short-term costs and benefits. When the Tennessee Valley Authority says it cannot afford to pay enough for coal to allow the damage done by strip mining to be repaired, it is attempting to pass the cost of its operation to the public in decreased environmental quality instead of increased power costs. Such a decision should be stated in the environmental statement in just this way, rather than viewing this type of economic impracticality as an obstacle to the adoption of alternative courses of action.

Technological impracticality must be viewed not only in relation to cost, which is well understood, but also in relation to time. Given enough time (and enough money), technological alternatives can probably be developed for most environmentally destructive actions. The question that must be asked is whether the need for a particular project is so immediate as to rule out the possibility of delay so that adverse environmental effects can be eliminated. For example, proponents of nuclear power plants raise the spectre of power shortages if such facilities are not constructed immediately. They tend to disregard the fact that the environmental consequences of such plants are very uncertain. The question becomes whether we Americans can change our power consumption habits enough to allow us to wait to eliminate harmful environmental effects before building nuclear plants. The answer depends on a whole series of economic, social, and political variables. Until we know whether these variables can be manipulated to allow time to develop alternatives, the claim of technological impracticality should be rejected.

703. Administrative and Judicial Interpretation

The performance of federal agencies in implementing the NEPA has been spotty, but seems to be becoming more consistent. The Department of Transportation did not issue a statement on the environmental effects of the SST until several months after the vote on the appropriations request in the House of Representatives, but later did delay the extension of runways at New York's Kennedy Airport pending review of the environmental impact by the National Academy of Sciences. Under pressure from Congressman Dingell, the International Boundary Commission delayed approval of the use of chemical defoliants along the United States-Canadian border while it solicited the comments of the affected states (but not, curiously, the comments of the affected Canadian provinces).

In several recent judicial proceedings the NEPA has been cited as authority to deny actions where adverse environmental effects might result. In *Zabel* v. *Tabb*,[9] the owners of land underlying Boca Ciega Bay in Florida required the issuance of a permit by the United States Army Corps of Engineers to fill in eleven acres of tidelands for use as a commercial mobile trailer park. Several state agencies and about seven hundred concerned citizens filed protests with the Corps. The United States Fish and Wildlife Service also opposed the application for a dredge and fill permit because such construction "would have a distinctly harmful effect on the fish and wildlife resources of Boca Ciega Bay."

The Secretary of the Army denied the application on the grounds that issuance of the permit would cause irreparable damage to the fish and wildlife resources. The landholders then brought suit in federal district court asking that the Secretary of the Army be required to issue a permit. The district court, relying on past inter-pretations of the Rivers and Harbors Act, held that the Secretary was without authority to deny a permit for reasons other than the obstruction of navigation. However, on appeal to the United States Court of Appeals for the Fifth Circuit, the court held that the Fish and Wildlife Coordination Act, when read together with NEPA, did give the Secretary of the Army power to prohibit an action where the ecology was endangered, and upheld the denial of the permit. Although NEPA was not in existence at the time Zabel was denied a permit by the Secretary, the court said that judicial review of an administrative decision must be made in terms of the applicable standards at the time of the court's decision.

In the *Wilderness Society* v. *Hickel* case,[10] the plaintiffs were the Wilderness Society, Friends of the Earth, and the Environmental Defense Fund, Inc., who argued that Walter J. Hickel, then Secretary of the Interior, should not be allowed to issue permits for the construction of the 789 mile long oil pipeline from Prudhoe Bay to Valdez, Alaska because the Trans-Alaska Pipeline System (TAPS) threatened the wilderness ecosystem of Alaska's North Slope. Specifically, it was claimed that the pipeline would cross high-risk earthquake terrain, and that there was a strong possibility that the pipeline would rupture and spill oil, causing irreparable damage to the environment. Second, the heated oil passing through the pipeline could irreparably melt and erode Alaska's delicate permafrost. Third, the pipeline road would use from 12 to 20 million cubic yards of gravel which would be taken from the rivers and streams of the public domain. Fourth, oil spills from tanker loading operations at Valdez would irreparably damage the Pacific Coast from Alaska to Seattle. An injunction was granted on the basis that Secretary Hickel had filed an environmental statement under NEPA which took into account road construction, but he did not consider the total environmental impact of the pipeline construction which would accompany the road. The injunction is important because it reverses the more common approach which is to issue licenses contingent on future compliance with vague standards. Projects of the size of TAPS gain great momentum and once initial construction is undertaken they are difficult to stop no matter what detrimental environmental effects may occur. For example, despite proven geological faults in the Santa Barbara Channel, and despite the occurrence of one disastrous oil spill, oil companies are still permitted to drill new wells in that area.

NEPA has also been used with some success in conjunction with other environmental protection statutes. The Environmental Defense Fund filed a petition with the Department of HEW in April of 1970 to force the Secretary to set emission standards for gasoline to implement the Clean Air Act and NEPA. The Clean Air Act itself does not require standards to be set for air pollution, but in combination with NEPA it can be argued to so require. Similarly, the Environmental Defense Fund filed a petition with the Federal Aviation Administration in May of 1970 to require the immediate setting of environmental standards to apply to the SST, citing NEPA and the Aircraft Noise Abatement Act as authorities. EDF requested standards for sideline noise, sonic booms, passenger radiation, and atmospheric pollution, claiming that a lack of standards would make it impossible to set later standards which would restore environmental quality as envisioned in the policy statement of NEPA.

In each of these proceedings, there is evidence that NEPA is beginning to affect the decisions of governmental agencies that have in the past been accused of being insensitive to environmental factors. For example the Atomic Energy Commission, which had always claimed that it had no power to consider environmental factors other than radiation damage resulting from its operations, has now instituted new procedures requiring license applicants to supply an environmental impact statement prior to public hearings for nuclear power reactor licenses. The Army Corps of Engineers have begun studying alternative ways to complete the controversial Cross-Florida Barge Canal with the objective of leaving more of the Oklawaha River free of dams. And, the Department of Transportation has rejected an application for a bridge permit where the proposed bridge would have led to the destruction of certain historic sites, thus having the sort of adverse environmental effect prohibited by NEPA. As federal agencies become more familiar with the substantive and procedural requirements of NEPA and their effect on statutory authority, it is to be hoped that they will increasingly incorporate this national policy into their decisions.

704. NEPA and Tort Law

Section 101(c) of NEPA provides that: "The Congress recognizes that each person should enjoy a healthful environment and that each person has a responsibility to contribute to the preservation and enhancement of the environment." This section and the responsibility of the federal judiciary to further the national policy suggest that the Act might have ramifications in the area of private tort law which currently offers the private citizen his only method of redress against polluters.

As indicated in the previous chapter, the present structure of tort law upon which NEPA is superimposed is one in which a plaintiff in a pollution abatement suit must base liability on one or more of the traditional doctrines of nuisance, trespass, negligence, or on a doctrine of strict liability such as products liability or liability for abnormally dangerous activities. Although only negligence among these requires a finding of fault on the part of the polluter, each calls for a process of weighing conflicting interests.

With few exceptions, the balancing of equities involved in the weighing process has treated primarily economic considerations, and has not taken into account the fact that a plaintiff's personal aesthetic interest in a clean environment is paralleled by the community's interest in abating pollution, and that this individual and group interest often outweighs the social utility of a polluter's contribution to the local economy. Now that Congress has articulated for the first time in the Act a recognition of each person's fundamental right to a healthful environment, it may be that the guarantee of the continued enjoyment of this right will weigh more heavily than before when it comes time for the court to balance conflicting interests.[11]

The narrowest impact which NEPA might have is to provide a basis for the abrogation of some common law defenses such as the polluter's acquiring over time the right to maintain a private nuisance or a trespass, and the victim's "coming to the nuisance" barring or at least making more difficult the recovery of damages. A broader impact would be the abrogation of some of the technicalities which make it difficult for a victim to make a case; such as the need for the victim in a public nuisance action to show special damages, or the rule that trespass occurs only where there are direct and not consequential damages. These implications arise directly from the recognition in section 101(c) of the Act that enjoyment of the right to a healthful environment is dependent on others respecting their duty to "contribute to the preservation and enhancement of the environment."[12]

705. NEPA and State Laws

A very broad effect on state law could be achieved under the theory that section 101(c) of NEPA sets up a federal right and duty which states must effectuate through their own systems of laws and provisions for private actions and remedies. If this argument were aimed only at overcoming the common law defenses and rules which hamper victims of pollution within the framework of the traditional common law remedies, then the resulting impact on state tort law and on the internalizing of costs of pollution would be substantial. If the argument were broadened to require the states to create new causes of action similar to NEPA the effect on state law and on effective cost internalization would be far greater, as the private victim would no longer be forced to sue under all the existing rubrics which private law has evolved.

A number of states have to date produced statutes which parallel or in some cases exceed NEPA in providing remedies. The Illinois General Assembly passed an Environmental Protection Act which established a Pollution Control Board, an Environmental Protection Agency, and an Institute for Environmental Quality. The Institute's function parallels that of the Council on Environmental Quality established under Title II of NEPA: it will carry out research and work with other state agencies in search of long-range solutions to environmental problems. The Pollution Control Board can promulgate standards for the control of air, water, and land pollution, refuse disposal, noise, and atomic radiation. The Agency may file complaints with the Board alleging violation of the Act or of a regulation, after which the polluter has the burden of showing that compliance with the law would create an "arbitrary or unreasonable hardship." The Board can issue cease and desist orders, impose money penalties, and grant individual variances.

In New York, an Environmental Conservation Law went into effect on June 1st, 1970 which declares a state environmental policy and creates a Department of Environmental Conservation to manage the environmental effort. The Department is the New York counterpart of the CEQ. The New York law goes beyond NEPA in that it empowers the Commissioner of Environmental Conservation to bring summary action against any person engaging in an activity which presents an imminent danger to the environment. Statutes similar to New York's have been passed in Connecticut, Illinois, and Kentucky.

Probably the toughest state environmental statute in the country and the one best suited to local conditions is the Vermont Environmental Protection Act passed in late 1970, which has been used as a basis for almost thirty individual actions against polluters in its first six months of existence. Vermont's environmental problem is acute because the state is used as a recreation area by city dwellers from Boston to Washington. This has produced both a Florida-type land boom in recreation land, and a general disregard for environmental protection on the part of these transient residents. Land developers have in many cases subdivided land into half- and quarter-acre lots with no provisions for central sewage. In some cases these developments are at altitudes of over 2500 feet where the ability of the land to absorb waste is low because of underlying bedrock. Where the soil cover is shallow, septic tanks overflow and wastes seep downhill into wells, streams, and lakes. Also large-scale building over 2500 feet, such as occurs with the development of ski areas, upsets fragile watersheds and other ecological balances, thus destroying fish and other wildlife.

The major provision of the Vermont statute is the establishment of a state environmental board and nine district commissions to establish comprehensive state capability, development, and land-use plans. A state-wide zoning law was established for all developments over 10 acres, with all construction for any purpose above 2500 feet altitude included. All included developments must be licensed and approved by one of the district commissions, which are empowered to hold public hearings and compel attendance of witnesses and production of evidence. An interesting feature is that there is no exemption for municipal development projects such as schools, waterworks, or industrial parks.

Another provision allows the state to buy land while still allowing farmers to work it. This allows the state to block a development without imposing economic hardships on existing farmers because of high taxes. This provision was needed because when a development comes into an area land values increase, and land becomes assessed at its market value rather than its value as farm land. Farmers are thus often forced to sell to the developers. In Windham County in Vermont, which includes many ski areas as well as a 23,000 acre International Paper development, an owner of a 55 acre wooded lot who paid $24 in taxes in 1967 would pay $690 in 1972. Each of these new laws, particularly those tailored to local problems as is Vermont's, helps assure compliance with NEPA while relieving conservationists of the costs and attendant perils of suing the federal government for compliance.

While state enactment of environmental protection laws is hopeful, the most promising approach to pollution abatement is on a worldwide basis, an approach which will be discussed briefly in Chapter Eight.

REFERENCES

[1]From President Nixon's environmental message to Congress, 1971

[2]Public law 91-190, 91st Congress, 1st Session (January 1, 1970), 42 U.S.C. 4331 et seq.

[3]For example, the Air Quality Act states as a Congressional finding "that the prevention of air pollution at its source is the primary responsibility of states and local governments."

[4]The structure and part of the content of the first two sections of this chapter were inspired by an excellent paper entitled Title I of the National Environmental Policy Act of 1969, by Ronald C. Peterson, a third year student at Yale Law School, which was written during 1970 while he was in residence at the Center for Law and Social Policy in Washington, D.C. under Yale's intensive semester program.

[5]Congressional Record, No. 115, Senate 17451 (December 20, 1969).

[6]In practice, the "action-forcing" requirement of a reviewable record has not worked out uniformly well. The language of NEPA seems to warrant the assumption that environmental statements must be circulated and made public before an agency decides to act. Yet the Act does not say positively when the statements are to be submitted or made public. Russell E. Train, Chairman of the White House Council on Environmental Quality, issued guidelines which indicated that a draft environmental statement should be prepared and circulated for comment to other agencies, but that only the final completed statement need be made public. However, some agencies have decided that "timely" publication is months after they have proposed an action, and even months after they have gone ahead with it. Thus, William M. Magruder, head of SST development, did not submit a draft impact statement to other agencies and the public until nine months after the Administration first asked for $290 million for the SST program, and did not make public the final impact statement and comments until after the House had first voted on the bill.

[7]The only major federal actions exempted from the requirement that a detailed statement be prepared are some which affect water quality. Under the Water Quality Improvement Act of 1970, a certification procedure was set up to secure advance compliance with water quality

standards. When compliance with water quality standards is secured from an appropriate state or interstate agency, NEPA statement filing requirements do not apply. However, exemption under this procedural requirement does not exempt water projects from other duties under NEPA where compliance with standards is required.

[8]A number of enlightening examples are given in a fascinating article to which I am indebted for a number of insights: Gene Marine, "The Engineering Mentality," *Project Survival* (Chicago: HMH Publishing Company, 1971), pp. 205-220.

[9]*Zabel* v. *Tabb*, 430 F. 2d 199 (5th Cir. 1970).

[10]*Wilderness Society* v. *Hickel*, Civ. No. 728-70 (D.D.C. April 23, 1970), *Environment Law Digest*, Vol. 70 (1970), p. 1.

[11]It should be noted that the Act in its final form states only that each person *should* enjoy a healthful environment; the legislative history is ambiguous as to whether this means that each individual has the "right" to a healthful environment, with the power to enforce it. In the original Senate bill the section contained somewhat stronger language referring to an inalienable right to a healthful environment, but this wording was stricken in the compromise with the House version of the Act. However, the Act does speak in positive terms of the "responsibility" of each person to contribute to the enhancement of the environment, and the section by section analysis retains strong language about the importance of a healthful environment even though at no point does it call it a right.

[12]These and other impacts on the technicalities of tort law are discussed in great detail in Virginia Coleman, "Possible Repercussions of the National Environmental Policy Act of 1969 on the Private Law Governing Pollution Abatement Suits," *Natural Resources Lawyer*, Vol. III, pp. 647-693.

8

The

Stockholm

Conference

Could there, one wonders, be any undertaking
better designed to meet the world's needs,
to relieve the great convulsions of anxiety
and ingrained hostility that now rack
international society, than a major international effort to restore the hope, the
beauty, and the salubriousness of the natural
environment in which man has his being?

George F. Kennan[1]

800. Introduction

Stockholm, Sweden was host to the first United
Nations Conference on the Human Environment, from June 5th
to 16th, 1972, which brought together almost a complete cross-
section of the world's 3.5 billion people with delegates from 114
nations. (The only major industrialized nation which boycotted the
conference was the Soviet Union.) The conference represented
what many saw as the beginning of a new era of international
collaboration to improve the earth's deteriorating quality of life.

The conference produced agreement, in principle, that nations,
despite their sovereignty, have mutual responsibilities for such
common property as the oceans and the atmosphere, and have
responsibilities to each other for constructive environmental
efforts. The nations were unanimous in their acknowledgement that
a worldwide environmental emergency existed, in sectors ranging
from urban blight to insecticide pollution, and that concerted international action was required. The meetings produced a 200-point

249

program of international action, designated a permanent organization within the United Nations to coordinate these actions, and adopted a code of principles to serve as guidelines for future national performance. The Conference's conclusions are pending ratification by the United Nations General Assembly.

Recognition of the need for internationally coordinated treatment of environmental pollution occurred precipitously. In 1967, when Sweden's U.N. delegates first urged an international meeting, the response was indifference. In 1969, when the issue was raised again, the General Assembly voted almost unanimously for the two-week debate in Stockholm. U Thant, then Secretary-General of the U.N., chose Maurice Strong, a Canadian industrialist as secretary-general of the conference. Strong turned out to be an ombudsman for the developing countries, who now numerically dominate the U.N. but who were at first reluctant to support the conference. They pointed out that the environmental problem facing the Third World nations was poverty; and that pollution was largely a disease of affluence.

In 1970, Strong requested that every country in the U.N. make a survey of national environmental conditions and problems, and submit a report prior to the conference. Two hundred international organizations, 20 U.N. agencies, and 100 internationally-known scientists were also asked for their ideas on the state of the environment. This pre-conference request forced countries of the world to set up a committee, involve their responsible people, and, in many cases, make their first surveys of the resources, needs, the state of pollution, the physical and social conditions in urban and rural settlements, plant and animal life, and their views on useful approaches to the environmental problem.

Strong then assembled experts from the U.N. and the Smithsonian Institution to evaluate the reports and to formulate six action plans on cities, natural resources, pollutants, information, economic aspects of environment, and the structure of an organization for continuing the work after the conference. Intergovernmental groups under a 27-nation Preparatory Committee were asked to draft concrete proposals on marine pollution, soil conservation, monitoring and resource conservation, and on a Declaration on the Human Environment. The document that emerged from this work, some 900 pages in length, was distributed to all participating nations.

"Mini-Stockholms" also were held in Rensselaerville, New York, on organization questions; in Founex, Switzerland, on problems of the developing countries; and in Canberra, Australia, on scientific monitoring problems.

The Founex conference, in June of 1971, was of special interest in that it focused the issues that would confront the industrialized countries in Stockholm. The report from Founex clearly stated that in the view of the majority of the participants, "dire poverty is the most important aspect of the problems which afflict the environment of the majority of mankind." The environmental problems listed by developing countries were: poor water supplies, inadequate sewerage, sickness, nutritional deficiency, and bad housing— all of which were aggravated by rapid population growth. Further, these countries stated their fear that if environmental measures increased the price of goods produced in the rich countries, the rich might try to impose protective taxes on imports from countries with less rigorous standards. The poor countries were also afraid that some exports, such as lead and high-sulfur fuel, would be displaced by nonpolluting technology; that recycling would reduce the demand for raw materials; and that fruits and vegetables containing DDT and similar substances would be banned. There was an expressed need that some early-warning system be worked out to inform developing countries of such pending advancements.

The pre-conference documentation in Founex urged all nations to limit the growth of their demand for energy, and ignored the fact that much of this growth is due to the replacement of human labor by electric power as part of the industrialization process. The effort to reduce industrial energy consumption thus ran headlong into the desire of the developing countries for higher levels of production. Since energy generation is never free of environmental impact (if only through heat emissions), the result was both an impasse, and the dropping of the demand for limited growth in energy production from the Stockholm sessions. Finally, there was apprehension in Founex that the cost to the developed nations of environmental protection would leave less money for already diminishing foreign aid.

Perhaps as a result of the meticulous pre-conference planning, the Stockholm sessions themselves were unexpectedly productive. The general accomplishments of the sessions were summarized by Maurice Strong in an interview with the Swedish press:

> . . . we set up machinery to work toward short and long range goals. To take issue with the technical questions, like who actually does the polluting in a given situation, there will be a monitoring

system, as part of a global assessment program. And there will be exchange of information and research programs on such questions as, for example, the encroachment of the deserts, substitutes for DDT, better methods of waste disposal, recycling, new clean technologies. And of course there will be measures proposed on problems like water and soil contamination—so crucial for developing countries, some of whom lose more capital each year through soil contamination than they receive in foreign aid. As for the depletion of plant life, where, for example, the practice of monoculture in relation to the green revolution is extinguishing many types of plant life which it took millions of years to evolve, we propose genetic banks to preserve them.[2]

Specific results of the conference included the approval of an "action program" involving 200 recommendations in fields that ranged from monitoring climate change or oceanic pollution to promoting birth control and the preservation of the world's vanishing diversity of plant and animal species. An Environmental Fund was approved to cover that part of the international effort not paid by specialized agencies and national governments. Those pledges which were made suggest that the fund will reach about $100 million, which is considered the minimum requirement for the first five years of operation.

An interesting series of conference recommendations were aimed at fears that the rapid adoption by farmers of standard crops is weakening the gene pool on which the long-term survival of such crops is dependent. Where a single strain or group of strains is used over a wide geographic area, the crops are vulnerable to blights against which they have no defenses. It is usually in wild or exotic strains that blight-resistant properties are found—strains which are fast vanishing. Thus, a recommendation of the conference was that governments initiate emergency programs to explore and collect species which are imperiled, and to store seeds and otherwise preserve and develop breeds for posterity.

One proposal that pitted the Third World nations against the industrialized nations was a proposal by India and Libya that an international fund or financial institution be set up to provide seed capital to help developing nations improve their housing. The Third World members easily passed this plan, but the 15 nations voting against it included virtually all of the more affluent countries who were expected to contribute to the plan.

801. The National Reports

The reports submitted by 78 nations on the state of the environment in their particular country produces both a sense of the immensity and diversity of the worldwide environmental problem as well as a feeling for the honesty and dedication with which solutions are being sought. The first report, by Afghanistan, concludes that:

> The population explosion, depletion of natural resources, inadequate waste disposal systems, unavailability of adequate potable water, the widespread introduction of DDT into the ecosystems, the countryside destruction of natural pastures by overgrazing and conditions related to the underdeveloped status of the country are the most important ecological problems in Afghanistan.

India submitted the most comprehensive survey, contained in four lengthy volumes. While for India population and poverty are the most crucial environmental problems, much of the Indian report discussed the fate of the country's wildlife. The report cited the increase in human population, which has upset the delicate balance of nature, producing species that are injurious to man and his products which are increasing in numbers as a result of man's slaughter of their natural predators. According to the report, "In India it is a case of the giants being superseded by the pygmies. The wild elephant, the rhinoceros, and wild buffalo, the lion, the tiger, and the wild bear, the black buck, the musk deer, the nilgai and the wild ass, the mongoose, the fox and wild cat, the golden eagle, the pink headed duck, the peacock, the Great Indian bustard, the florican, the quail and the partridge either have become extinct or are fast disappearing. What have multipled are the mosquitoes and sand fleas, ticks and mites, the caterpillars, moths, locusts, ants, beetles, and insects of all sorts which attack crops, the sparrows and the saragas, the monkeys and the langurs, the mice and rats which do great damage to the crops and seem to be yearly increasing in numbers and in their ravages."

Holland's report indicates that it faces perhaps the most crucial environmental problem in Europe. Not only is Holland the most densely populated country in Europe, with 350 people per square kilometer, but the problems of effluents and waste disposal are aggravated by increasing industrialization and decreasing supplies of fresh water. The low elevation of the country means the constant danger not only of flooding, but also of salinification of scarce surface and ground waters, and of permeation of the soil by salt water. There is a tremendous need for fresh water to counteract this—for this fresh water, Holland is dependent on the Rhine and the Meuse rivers. But the salinity of the Rhine is increasing yearly due to waste salts discharged from the potash mines in Alsace and coal mine drainage in the Ruhr. Holland also faces the problem of pollution of the North Sea, the Dutch beaches, and other coastal areas through the discharge of oil and wastes.

Kenya submitted a report which stated: "There is a danger that the people of Kenya, as the peoples of developed industrial nations have done in relation to the toxic wastes of their factories, will adapt to a steadily deteriorating environment. People in certain parts of Kenya, for instance, do not see soil erosion as a problem but rather as a basic feature of their environment to which they adapt as best they can." The Kenyan report also states: "If present trends continue we can expect an increasing national bill for famine relief; a steady decline in the options open to us in developing resources not as yet fully utilized such as wildlife, domestic livestock, scenic attractions and the harnessing of water supplies; land use conflicts exacerbated by unplanned population pressures from a nation expecting a piece of land for everyone; deteriorating quality and quantities of water; and the total destruction of many delicate and valuable ecosystems."

China submitted a brief report in which it described not the problems of the Chinese environment, but rather the Chinese approach to pollution control. The report states: "There are three wastes, being air, water, and residual waste, and three benefits, being the recycling of these resources for reuse. No industrial firm can open its doors in China until it has satisfied the above motto. The Chinese people have decentralized industry into rural areas, and use human and animal waste as feed and fertilizers."

The DDR (East Germany) submitted a report reflecting the same pollution problems as other industrialized countries, but emphasizing the problem of diminishing recreational areas. "In 1949, there were 350,000 vacationers, compared to over 9.5 million today, thanks to the five-day work week. Thus, a broad expansion of recreational areas is in progress. Due to the lack of coastal area, artificial lakes and pools are being constructed inland, with facilities for camping and cottage living."

West Germany included in their discussion of domestic problems those of the job environment, where "twenty-seven million people are employed and are thus exposed for about eight hours per day to the special effects caused by the work process. Noise, vibrations, heat, cold, polluted air, vapors or other substances harmful to human health are just a few examples of the effects of the job environment on man." Among new measures to improve the job environment in Germany is a law that requires that each place of work must be arranged so that dangers to people's environment are minimized or excluded.

802. What Was Left Out?

For all its comprehensive coverage of the problems of the world environment, several basic issues were not included in the Stockholm Conference. Because of the political character of the conference, the obvious problem of the control of population growth was not raised. The prime issue of birth control was dismissed as a thorny matter best left to a scheduled 1974 United Nations conference on population. Also, what is perhaps the most immediate threat to the human environment—the everpresent threat of thermonuclear war—was not mentioned even once in the entire 900 pages of conference documentation. The use of chemical defoliants and similar materials in Vietnam was raised and discussed, but no important conclusions were reached and no strong resolutions were adopted.

In conclusion, and in spite of its shortcomings, the Stockholm conference was probably more important for the change in national attitudes that it symbolized than for what it produced. The anthropologist, Margaret Mead, invited to address the conference on behalf of the nongovernmental observers in attendance, said:

This is a revolution in thought fully comparable to the Copernican revolution by which, four centuries ago, men were compelled to revise their whole sense of the earth's place in the biosphere. Our survival in a world that continues to be worth inhabiting depends upon translating this new perception into relevant principles and concrete action.

REFERENCES

[1]George F. Kennan, in *Foreign Affairs* (1970), quoted in *Conservation Foundation Newsletter* (November, 1971), p. 3.

[2]Ruth Link, "How Do We Want Our World," *Sweden Today*, Vol. 6 (1972), p. 36.

Glossary

Abatement. With regard to pollution and polluters, abatement means the attempt to diminish or remove the pollution that is causing the immediate problem. When faced with an unacceptable degree of air pollution, the mayor of Pittsburgh may order a halt to the burning of high-sulphur coal, or may consider closing off some heavily traveled streets to automobile traffic.

Abiotic Environment. The interaction of physical and chemical and inorganic ingredients in a specific environment.

Aerator. A mechanical device used to inject air into waste-containing water in order to facilitate aerobic action.

Aerobe. An organism that requires oxygen in order to survive.

Aerobic. The action of bacteria utilizing free oxygen. Also refers to the presence of oxygen in air or water or to organisms which require oxygen for survival.

Aerobic Waste Treatment. A method of subjecting waste-containing water to aeration, with the wastes being partly purified by oxidation.

Allocative Efficiency. An economic criterion that allocates resources in a way that results in maximum efficiency for society. A necessary condition for the optimal allocation of resources is that the marginal product of any resource be the same in all its alternative uses. With pure competition in product and resource markets an optimal allocation is achieved automatically. However, monopoly or monopsony or the existence of external costs or benefits may lead to misallocation.

Alternative Costs. (See Opportunity Costs.)

Ambient Air. The air "outdoors," extending from ground level to about 10 miles in altitude, or that air which is subjected to meteorological and climatic changes.

Amortization (Of Fixed Assets). A method of depreciating the original investment in equipment or plant facilities over the estimated average service life of the asset.

Anadromous Fish. Fish which grow and live in the sea, but migrate up a river to spawn.

Anaerobic. The action of bacteria not utilizing free oxygen. Also refers to the lack of oxygen in a particular environment or to organisms such as certain bacteria which do not require oxygen for survival.

Angstrom. A unit of measurement, being one ten-thousandth of a micron, which is used to measure radiation.

Bayesian Statistics. A method of statistics which uses the best subjective estimate of a given circumstance as if it were a firm probability (instead of using firm probabilities).

Benthic. All animal or plant life inhabiting the sea or lake bottom.

Beryllium. A metallic element, which is an ingredient of rocket fuels and which is also given off in exhaust gases from some industrial processes.

Biota. The collective plant and animal life found in a specific region or environment.

Biotic Environment. The interaction of plants, animals, and microorganisms in a specific environment.

Biotic Potential. The potential of a living species to grow in size. Some forms of pollution can limit the biotic potential of any species if the quantity and time exposure of the pollutant is sufficient.

Capital Investments. Investments in capital goods which are used in the production of other goods; such goods include factory buildings, machinery, trucks, and tractors. Land and money are not usually considered as capital goods.

Carbonaceous Oxygen Demand (COD). A measurement of the amount of oxygen in parts per million required to oxidize organic and inorganic compounds in water.

Carboxyhemoglobin. A compound which forms when blood combines with carbon monoxide. This hemoglobin compound is stronger than blood forms with oxygen. The stronger compound is called carboxyhemoglobin (COHb). Normal COHb levels in the blood are about 2 percent; a heavy cigarette smoker may have a level of 5 percent. A level of 10 percent, which could occur with combined air pollution and heavy smoking, would likely cause death.

Carcinogens. Cancer-producing substances.

Caveat Emptor. "Let the buyer beware."

Caveat Venditor. "Let the seller of goods beware" (or be responsible for defects or deficiencies of the goods).

Chemosterilants. Chemical compounds used as sterilizing agents.

Class Action. A legal action brought on behalf of other persons similarly situated.

Class Representation. Where members of a class sue or are sued on behalf of other members, judgment is conclusive for and against those members of class who are represented except where there is fraud or collusion.

Coliform Bacteria. Nonpathogenic microorganisms used to indicate the possible presence of pathogens in water. Coliform bacteria enters water from fecal deposits of humans or animals.

Coliform Count. A count of coliform bacteria used to indicate the presence of pathogens. If the coliform count is greater than zero, the water is unsafe for drinking but possibly useful for other purposes such as fishing and boating.

Colloids. Matter suspended in other matter, with particle sizes ranging from 10 to 1000 angstroms, is classified as being in a colloidal state. Colloidal matter cannot settle out of the circulating medium through the force of gravity.

Composite Index. A measure of the relative changes occurring in a series of values compared with a base point. The base period usually equals 100, and any changes from it represent percentages. The index may be constructed of several values; for example, an air pollution index may combine values for sulphur oxides and particulates present in the air at any one time.

Constant (Value) Dollars. A series of dollar values such as gross national product, personal income, or profits, from which the effect of inflation in the value of the dollar has been removed. The resulting series is in real terms rather than in dollar terms, and thus is a better measure of physical volume. The process of converting current dollar values into constant dollar values involves deflating the former figure by some factor related to the value of the consumer price index (or wholesale price index) to equivalent dollars in some preselected year such as 1940, 1950, 1960, etc.

Convection. The process of heat transfer through the movement of air masses or of water masses.

Correlation. A statistical technique which relates a dependent variable to one or more independent variables over a period of time to determine their closeness.

Cost-Benefit Analysis. A method of analysis which enumerates and evaluates all the relevant costs and benefits associated with a project. If the total of discounted net benefits exceeds the capital cost being contemplated, then the project is economically reasonable and is put in a pile with other feasible projects to be ranked according to some other criteria, usually either the ratio of benefit to cost, or else purely political considerations. If discounted benefits are less than anticipated costs the project should be rejected on purely economic grounds.

Cross Elasticity. A measure of the influence of the price of one good on the demand for another. The cross elasticity of demand for good "x" in terms of the price of good "y" is the percentage change in the quantity bought of "x" divided by the percentage change in the price of "y," with consumer tastes, money incomes, and all other prices being constant. Cross elasticity measures the degree of relationship between two goods; the higher the cross elasticity between "x" and "y," the greater their interdependence.

Crustaceans. A class of small marine life, including shrimp and lobster.

DDD. One of the decay products of the insecticide DDT.

DDE. One of the decay products of the insecticide DDT.

DDT. A persistent pesticide, the chemical name of which is dichlorodiphenyltrichloroethane. DDT is also known by the names Arkotine, Dicophane, Gerasol, and others.

Degradation. The breakdown of large organic molecules into smaller molecules producing inert material on the one hand and stabilized material on the other.

Diffusion. The process of spreading heat, light, gases, or liquids. It occurs from internal molecular movement or from external factors such as atmospheric turbulence.

Diffusion Model. An attempt to model and predict patterns of diffusion.

Dust Concentration. Industrial or natural particulates, 76 microns or less in size, are called dust. A micron is 3.5×10^{-5} inches. Any dust smaller than 5 microns can pass through the nostrils and lodge in the alveoli of the lungs.

Dysbiotic. A chemical that can create genetic defects in plants, animals, or microorganisms is said to be dysbiotic.

Econometrics. The area of economics which expresses economic relationships in mathematical terms in order to verify them by statistical methods. Econometrics attempts to measure the impact of one economic variable on another to predict future events or advise the choice of economic policy to achieve desired results.

Ecosphere. The biological sphere that contains life on earth, in the oceans, and in the air; it contains the various levels of atmosphere and all living things therein.

Ecosystem. The relationships between living things and their supporting earth environment.

Effluent. Gases or liquids containing pollutants, which are expelled into the environment.

Effluent Fees. Fees paid by industrial polluters for the right to continue polluting, usually based on the damage done by their discharge, or on the cost of cleaning up the pollution at some central point.

Electrostatic Precipitator. An air pollution control device that removes particulates from a gas stream. The gas is passed through a high voltage field, whereby its molecules are ionized and the charged particles attracted to collecting electrodes. The collected particles are then mechanically collected while the clean gas is discharged into the air.

Endometrial Cancer. Cancer of the endometrium (the mucous membrane lining of the uterus).

Epidemic. Usually refers to large scale deaths caused by water-borne or air borne bacteria, or by rat-carried viruses. Multiple deaths attributed to pollution are more commonly called episodes.

Erosion. Erosion occurs from weather, runoff, or scour, depositing quantities of silt in surface waterways. Man-made erosion occurs when man cements natural surfaces used in normal runoff. As a result water has to carve out new paths to arrive at its destination, thus causing erosion.

Estuaries. Zones along the coastline that are areas of interchange between land drainage and ocean tides. These are particularly important as areas where ocean fish spawn.

Eutrophication. A cycle whereby algae feed upon some source of nutrients (for example, nitrogen and phosphorous from sewage and detergents), and dead and decaying algae consume so much oxygen from the water that bacterial action which would have broken down organic wastes is destroyed. At some point in a natural lake or pond the dissolved oxygen level becomes so low, the oxygen demand becomes so high, and algae growth is so acute that the body of water reaches the point where it is about to "die."

External Costs. (See Externalities.)

Externalities. Costs or benefits which accrue indirectly to individuals or to society as a result of actions over which those people or society have no control. The cost of excess dry cleaning bills resulting from the smoke of a smelter falls on the residents of the community rather than on the owner of the smelter, who takes no account of these external costs in his own profit and loss calculations.

Fossil Fuels. Coal, oil, and natural gas are fossil fuels. About 75 percent of the air pollution in the United States stems directly or indirectly from the combustion of fossil fuels.

Genossenschaften. River basin associations in Germany.

Hydrocarbons. Substances composed of only carbon and hydrogen, an example of which is the octane found in gasoline. Hydrocarbons are made up of four main groups: paraffins, naphthenes, olefins, and aromatics.

Hydrolysis. A process whereby a salt reacts with water to change the pH of the water, the salt is hydrolyzed and the process is called hydrolysis.

Linear Programming. A mathematical technique designed to select from among alternative courses of action the one most likely to achieve a desired goal, such as achieving a given pollution reduction objective at the lowest possible cost.

Macroeconomics. Economic analysis that is concerned with data in aggregate as opposed to individual data; for example the analysis of the general price level rather than the prices of individual commodities, or of total employment rather than employment in an individual firm.

Marginal Cost. The additional cost that a producer incurs by making one additional unit of output. If, for example, total costs were $5,000 to produce two tractors per day and $8,000 to produce three tractors per day, the marginal cost of the third tractor is $3,000. A purely competitive firm, facing a given price set in the marketplace, will increase its output until marginal cost equals price.

Marginal Revenue. The additional revenue that a producer incurs by selling one additional unit of output. If, for example, total revenue were $6,000 from selling two tractors per day and $8,500 from selling three tractors per day, the marginal revenue of the third tractor is $2,500. A producer will expand his output until his marginal revenue equals the additional cost of producing an additional unit of output for sale.

mg/100 ml. Milligrams per 100 milliliters. For example, the normal range of lead in the blood is measured as from 0.015 to 0.045 mg./100 ml. of blood.

Microeconomics. Economic analysis that is concerned with data in individual form as opposed to aggregate data; for example, the study of the individual firm rather than aggregates of firms, or of the relative prices of individual goods rather than aggregate price levels in the economy.

Micron. A measure of size, a micron is 3.5×10^{-5} inches (or one ten-thousandth of a centimeter).

Molecular Sieves. Plates with minute holes that permit molecules of contaminated gases to be exposed to the absorbent in an absorption scrubbing process.

Molluscs (or Mollusks). A group of invertebrate animals which include oysters, clams, snails, and octopus.

Morbidity. The rate of illness.

Mortality. The death rate.

Multiple Correlation. A technique of statistical correlation, which uses more than one independent variable.

Multivariate Regression Analysis. Regression analysis which uses more than one independent variable.

Mutagenic. A chemical that can create inheritable genetic change in living creatures.

Nitrates. Nitrogen compounds found in water.

Normative Levels of Pollution. A normative level of pollution refers to an idealized level (perhaps zero pollution). Since no idealized standards do or can exist, there are no definitions of normative environmental conditions that have any moral superiority over others except by reference to the selfish needs of one portion of society.

Opportunity Costs. With a relatively fixed supply of labor and capital at any given time, the economy cannot produce all it wants of everything. Thus the real cost (the opportunity cost) to society of an opera, a school, or a dam is the value of other things that cannot be produced because the same resources are not available to build them. Opportunity costs are best illustrated in wartime when a nation must choose guns and tanks over new automobiles and appliances, but cannot have both.

Optimal Levels of Pollution. An optimal level of pollution (for example, air pollution) is one in which atmospheric composition contains some oxygen, some carbon dioxide, and some hydrogen sulfide and particulate matter, in concentrations which permit society to pursue the greatest possible satisfaction for its human members. An optimal state is best expressed in terms of the other goods such as housing, ballet, and medical care, that must be foregone in return for somewhat cleaner air.

Oxidation Lakes. Artificial lakes or lagoons into which industrial wastewater is piped; air is bubbled through the water to supply oxygen for the oxygenation process.

Ozone. A colorless, unstable, pungent gas produced as a by-product of the photochemical processes. It is found in the earth's atmosphere as O_3.

Pathogenic Bacteria. Microorganisms that cause health hazards.

Pathogens. Organisms that can cause diseases in man. Pathogens may be airborne, waterborne, or transferred from man to man or from animal to man.

Particulate Loading. The introduction of particulate matter into ambient air.

Particulates. Organic and inorganic particles, both observable and invisible, which are found in smoke emissions and in ambient air.

Permafrost. The subsoil in arctic regions which remains frozen throughout the year.

pH Value. A measure of acidity or alkalinity. The range of pH values for water are from zero to 14. Neutral water has a pH of 7; less than 7 reflects acidity, more than 7, alkalinity.

Prescriptive Rights (in law). To create an easement by prescription, the use of property must have been open, continuous, exclusive, and under claim of right for the appropriate statutory period. The term prescription is usually applied to incorporeal hereditaments, while "adverse possession" is applied to lands. For example, a prescriptive corporation is one which has existed beyond the memory of man, such as the city of London.

Regression Analysis. A statistical term used to indicate a method of measuring a relationship between two or more variables. A regression line (or least-squares line) is derived from a mathematical equation to relate one variable to another. For example, an increase in sulphur oxide concentrations in the air results in an increase of some kinds of bronchial tract disorders, and a regression line can be used to show this relationship.

Res ipsa loquitur. "The thing speaks for itself."

Somatic Cells. All the cells of the body which are not sexually reproductive—as distinguished from germ cells which are sexually reproductive.

Submicronic. Smaller than one micron, e.g., smaller than 3.5×10^{-5} inches.

Symbiosis. The interaction between two or more organisms whereby both thrive on their interdependence with the other.

Synergism. The effect of two or more separate substances such that the joint effect on the receptor is greater than the sum of the individual effects. In many cases the danger to health from two or more combined pollutants exceeds the sum of their individual dangers.

Temperature Inversion. A condition in the lower atmosphere whereby the temperature of rising air does not cool with an increase in elevation, causing a spread of polluted air horizontally rather than vertically.

Teratonogenic. A chemical that can create inheritable genetic change in lower forms of life.

Terracide. Refers to death of the earth by human hands.

Terricide. Refers to death of life in one specific part of the earth by human hands.

Tort. A private or civil wrong or injury which is not covered under a contract. It includes a direct invasion of some legal right of the individual, or the violation of some private obligation by which damage accrues to an individual.

Trophic Levels. The different levels of nutrition of organisms. Trophic levels can be visualized as a triangle with plants or producers forming the base and carnivores the apex. Herbivores or primary consumers feed on plants. In turn, carnivores or secondary consumers eat the herbivores. Finally, there are tertiary consumers, which are carnivores which feed on other carnivores.

Turbidity. The muddiness or cloudiness of a body of water.

Index